SpringerBriefs in Computer Science

T0213853

For further volumes:
http://www.springer.com/series/10028

Peng Lin • Xiaojun Feng • Qian Zhang

Auction Design for the Wireless Spectrum Market

Springer

Peng Lin
Department of Computer Science and Engineering
Hong Kong University of Science and Technology
Clear Water Bay, Hong Kong SAR

Xiaojun Feng
Department of Computer Science and Engineering
Hong Kong University of Science and Technology
Clear Water Bay, Hong Kong SAR

Qian Zhang
Department of Computer Science and Engineering
Hong Kong University of Science and Technology
Clear Water Bay, Hong Kong SAR

ISSN 2191-5768 ISSN 2191-5776 (electronic)
ISBN 978-3-319-06798-8 ISBN 978-3-319-06799-5 (eBook)
DOI 10.1007/978-3-319-06799-5
Springer Cham Heidelberg New York Dordrecht London

Library of Congress Control Number: 2014939297

Printed on acid-free paper

Springer is part of Springer Science+Business Media (www.springer.com)

To my co-supervisor Mounir Hamdi, my colleagues in Qian's group and Huawei-HKUST lab, and all those who are studying economic mechanisms in wireless networks area.

Preface

Auctions are popular trading mechanisms, where bidders (buyers) offer money for some product or service, usually referred to as a "lot". After going through a process of taking bids from members of an audience, the auctioneer (facilitator) awards the "lot" to the highest bidder.

Auctions have a long history which goes back as far as 500 B.C. and were a daily part of the economic lives of the Babylonians and Ancient Romans. Over the centuries, auctions have become more and more mature and sophisticated. Generally, auctions are recognized as a fair and efficient means of trading which give equal opportunities to both buyer and seller. They are also economically viable as the lots are awarded to those who bid the highest, which means the winner values the contested lot over the other bidders. This in turn means higher revenue for the seller and auctioneer.

However, we do not aim to make a comprehensive survey of auction theory, as there are already plenty of books and papers on the subject. Rather our emphasis is the field where auctions and wireless networks intersect, therefore after we give a preliminary introduction to auction theory, we then turn to wireless networks. The purpose of Chap. 1 is to present the fundamental knowledge of our study. It is suggested for the readers, especially non-professionals, to study this chapter to gain some understanding. We hope readers catch the two main points when reading this book. First, what makes the auctions different in wireless networks and in general scenarios? Second, how are new auction mechanisms for wireless networks designed? Chapter 2 answers the first question, and Chaps. 3–5 present three examples to show how we design auctions for new wireless scenarios.

Spectrum auction is widely applied in spectrum redistributions, especially under the dynamic spectrum management context. Different from traditional auctions, spectrum auctions possess unique properties, such as interference relationship, reusability, divisibility, composite effect and marginal effect, which have not yet been well studied in previous works. Due to the newly developed properties, auction mechanism has to be tailored to wireless spectrum scenarios.

In this book, we first introduce the auction mechanisms for the wireless spectrum market. Chapter 1 is written to ensure even the non-professional readers understand the basics of auction mechanism and wireless networks. Chapter 2 summarizes the current research process of spectrum auctions. We introduce many papers published in top conferences and journals from the year 2008 until 2013. We present their ideas, make comparisons and summarize their advantages and disadvantages. Indeed we can see the inherit relationships among these works. Then we discuss three special scenarios where we design new auction mechanisms. In Chap. 3, we design a new truthful double auction framework for spectrum trading considering both the heterogeneous propagation properties of spectrums of different frequencies and spatial reuse. In Chap. 4, we design a framework to enable spectrum group buying for secondary users with limited budgets. Our framework points out how to select the bidders inside a group, allocate channel capacity and charge them, and reasonably decide the group bid. In Chap. 5, we design a flexible auction mechanism for operators to purchase the right amounts of spectrum at the right price according to their dynamic demands. Flexauc can also be applied to cases where the marginal decreasing effect works for the bidder. By applying this, the auctioneer may be able to obtain the optimal allocation result before going through all the bids, which significantly reduces the computational overhead. Flexauc can even run in linear time.

Hong Kong
July 2013

Peng Lin
Xiaojun Feng
Qian Zhang

Acknowledgements

This book was supported in part by grants from RGC under the contracts CERG 623209 and 622410, HKUST grant SRFI11FYT01, the grant from Huawei-HKUST joint lab, and the National Natural Science Foundation of China under Grants No. 60933012.

Contents

Acronyms

DCP	Diverse clearing price
FCC	Federal Communications Commission
Flexauc	Flexible auction designed by the authors
MM	Maximum matching
MWM	Maximum weight matching
SAMU	Winner determination algorithm scarifying at least m users
SAP	Secondary access point
SH	Spectrum holder
SINR	Signal-to-interference-noise-ratio
SN	Secondary network
SU	Secondary user
TAHES	Truthful Double Auction for Heterogeneous Spectrums
TASG	Three-stage Auction Framework for Spectrum Group-buying
UCP	Uniform clearing price
VCG	Vickrey Clarke Groves auction
WSP	Wireless service provider

Chapter 1
Introduction

1.1 Wireless Communication Networks

Nowadays people live with a variety of wireless communication networks and devices. All the networks and devices, such as Wi-Fi networks, cellular networks, satellite TV and Blue-tooth microphones, work on the basis of wireless communication networks.

As early as year 1895, the Italian inventor Guglielmo Marconi worked on long distance radio transmission and telegraph systems. Since then, more and more wireless communication systems have been developed for the purpose of data transmission, voice service, geographical coverage, and the support of mobile communication and social networks. In the last two decades, the research process for wireless communication networks has again sped up greatly. The emergence of cognitive radio technology and software-defined radio have improved the utilization of spectrum bands such that there will be room for the rapidly developing wireless technologies.

Wireless networks are built on the basis of the radio wave transmission of dedicated frequency bands. Interference can occur when two sources in close proximity transmit simultaneously at the same frequency, meaning neither source can correctly decode the signals. Spectrum management is utilized so that different wireless applications work on different frequency bands and thus avoid harmful interference. Further they can use interference alignment and cancellation technologies to improve the performance.

1.2 Economic Incentive Mechanisms

With the rapid development of wireless communication networks in recent years, they have become highly diverse and complex. The design of systems and mechanisms for wireless communication networks is facing two challenges. The first

P. Lin et al., *Auction Design for the Wireless Spectrum Market*, SpringerBriefs in Computer Science, DOI 10.1007/978-3-319-06799-5_1, © The Author(s) 2014

is that spectrum frequency is a scarce resource with interference and reusability properties. It is also a key to allocating the spectrum such that it is fully utilized and generates the maximum social welfare. The second is that wireless networks usually involve multiple self-interested entities (i.e., individual users/operators, primary/secondary entities). There are often conflicting goals among the regulators, commercial operators and end users in the market. The regulator may be nonprofit and care most about fairness and the utilization efficiency of the precious spectrum resource. The commercial operators are profit-driven, which means they would like to reduce the spectrum and infrastructure cost and raise the prices of their services. Meanwhile, they have to compete on price with opponents and launch attractive new services to hold and increase their customer base. End users trade-off their budget and expected service quality and make a selection from these operators and services. How to balance the interests of all parties, is also an open problem.

Under such circumstances, economic incentive mechanisms play a more and more important role in the spectrum resource allocation problem. Indeed, an increasing number of research works have focused on the design of economic mechanisms in a variety of scenarios, that is, distributed control of network systems, revenue maximization with pricing policies, and the most common resource allocation. Researchers have reached a consensus that to better build, understand, maintain, optimize, and upgrade large distributed networks, it is important to design economic mechanisms as well as technologies. Right now wireless networks are also at the cutting edge of their evolution, where dynamic spectrum accessing and cognitive radio technologies bring economic and incentive issues to the fore. There are opportunities to build economic incentives into the network architecture and protocols under development, and avoid the many problems that have arisen in previous network system design due to their lack thereof.

Despite its importance, until several years ago there had been very few research works done on the problem. Since then, plenty of well-developed models and terms of the economic and operations research areas have been applied to wireless networks scenarios, to better study and understand the nature and influence of the conflicting interests, and to optimize the system, making it fairer and more efficient. Among these models and terms, game theory, a study of strategic decision making, is the most frequently used. However, also frequently used are pricing theory, auction theory, contract theory and optimization technique. Some problems are very similar to those in the economic field. However, more problems have their unique characteristics. For example, the reusability of spectrum in temporal/spatial/frequency domain makes it different from general commodities. Such that the original economic models should be tailored to wireless network scenarios. Furthermore, sophisticated frameworks, consisting of multiple simple models, can be created to solve difficult problems. These uniqueness makes the mechanism design challenging.

1.3 Auction Models in Spectrum Allocation

There are too many interesting economic mechanism design works in the wireless communication network scenarios. In this book, we focus on spectrum allocation with auction models. Since 2007, there have been more and more research works studying new mechanisms on spectrum auction, transplanting existing models and designing new ones to solve problems for various wireless scenarios. However we can see that in real world auctions, the adopted mechanism is much simpler than those in the research papers. We believe the reasons they give up elegant mechanisms and take up the simpler ones include:

1. In the research literature, many assumptions are made. For example, it is assumed that the bidders do not collude. However in the real world it is hard to identify whether or not some of the bidders collude to manipulate the auction results. It may also be assumed that the auctioneers have some prior knowledge about the buyers and sellers, for example, the distribution of their true valuations about the items and the interference relationships between them. However there may be new participants and the auction may be dynamic such that there is no prior information available.
2. In the research literature, a nice property which is gaining attention is truthfulness (strategy-proof), which means that the mechanism is designed such that no bidder has the incentive to submit false values of any items. However in real world, people do not care so much about this aspect, partially because truthfulness is built on the assumption of no collusion. Besides, by achieving truthfulness, the seller or auctioneer usually pays by the reduction in the revenue and social welfare.
3. Concerns about fairness. Vickrey Clarke Groves auction (VCG) payment mechanism is very decent in the mathematical form, but it charges bidders different unit prices even for the same items. It may be hard for bidders to accept the price discrimination.
4. Concerns about complexity. For example, the VCG payment mechanism with combinatorial auction comes with high computational complexity, exponential to the input size. In the real world, it may be a problem, especially for the short-term secondary market auctions.

Therefore, the academic community has still a long way to go to make the newly designed mechanisms accepted by the real world.

In this book we are honored to share our views and research outputs with readers from all disciplines and with diverse background knowledge. In order to make it popular and easy to understand, we have done much survey work in introducing the background, scenarios and problems themselves. Then we present three of our recent works in spectrum auction: auction design for spectrums with heterogeneous propagation property, spectrum group-buying framework and flexible auction framework.

In our first work, we notice that a key uniqueness in spectrum auction is that the spectrum resources can be reused in different locations. However, due to the heterogeneous propagation properties of spectrums of different frequencies, their reusability is different. To enable truthful double auctions of the spectrums, we need to consider such heterogeneity. Therefore, we design a new auction mechanism call TAHES. In TAHES, the spectrum reusability is modeled with conflict graphs and double auctions will be carried out based on the heterogeneous graph for different bands.

In our second work we aim to build a group-buying framework for spectrum trading. In the secondary spectrum market, users from secondary networks may ask for spectrum frequency. However, individual users with limited budgets cannot afford a whole spectrum block on their own. Inspired by the emerging group-buying services on the Internet, for example, Groupon, we propose that users can be voluntarily grouped together to acquire and share the whole spectrum band sold in the spectrum auctions. However, there are two major challenges. The first is how to deal with users' different evaluations and budgets such that the trade is valid; the second is about how to design a truthful (strategy-proof) and efficient mechanism. We give a three-stage auction-based framework for this problem.

In our third work, we study the spectrum auction in the primary market. In the wireless market, major operators buy spectrum through auctions held by spectrum regulators and serve end users. How much spectrum should individual operators buy and how should they set the optimal service tariff to maximize their own benefits are challenging and important research problems. On one hand, a wireless service provider (WSP)'s strategies in spectrum auction and service provision are coupled together. On the other hand, for spectrum holders, how to design auctions to flexibly satisfy operators' spectrum requirements, to improve sales revenue and maximize social welfare remains unsettled. Previous works have not looked at the big picture and usually studied one of the sub-problems. We jointly study the spectrum holder strategy in auction and the WSPs' strategies in service provisions. We point out the relationship between their optimal strategies. To meet the WSP's flexible requirements, we design a flexible auction scheme (Flexauc), a novel auction mechanism to enable WSPs to bid for a dynamic number of channels. We prove theoretically that Flexauc not only maximizes the social welfare but also preserves other desirable properties such as truthfulness and computational tractability.

1.4 Book Organization

There are several chapters in this book. First we introduce the wireless spectrum market and spectrum allocation in Chap. 1. Then in Chap. 2 we give a background survey of auction mechanisms and point out the special and unique characters of spectrum auctions. We also briefly go through the relevant typical spectrum auction works. We have three of our own works to present. In Chap. 3, we describe a truthful auction design for heterogeneous spectrums. In Chap. 4, we present a spectrum group-buying framework for the secondary market. In Chap. 5, we propose a demand-sensitive flexible spectrum auction model. Chapter 6 summarizes our works.

Chapter 2
Auction Mechanisms

2.1 Overview

According to Wikipedia [1], "An auction is a process of buying and selling goods or services by offering them up for bid, taking bids, and then selling the item to the highest bidder. In economic theory, an auction may refer to any mechanism or set of trading rules for exchange."

There are some common auction forms used in the real world. The English auction is known as the open ascending pricing auction. During a sale an auctioneer calls out a low price and raises it until there is only one interested buyer remaining. The Dutch auction is known as the open descending price auction. The auctioneer calls out a high enough price such that there is no one interested in buying it. Then he gradually lowers the price until one buyer is interested. The price information is made public in both of these types of auctions. There are also sealed-bid auctions where bidders submit their bids and bidders are only aware of their own value. The auctioneer selects those who offer the highest bids as winners. The rule of winners' payment can be flexible. If a winner pays the second-highest bid rather than his own value, then the auction is a Vickrey auction (for a single-object auction).

In this book, to align with the definitions in most of the research works, we assume the spectrum auction is seal-bid.

Generally, there are three types of players in an auction: buyers, sellers, and the auctioneer. If these players are all present, it is a double auction. When the auction is held by a single seller (or buyer), he may act as auctioneer. We assume the buyers and sellers are selfish and rational, which means they behave within the auction framework and try to maximize their individual utilities. They do not submit true values if it brings them any extra benefits. We also assume the auctioneer faithfully executes his duty. Therefore when a seller (or buyer) acts as auctioneer, all participants unconditionally trust him.

An auction consists of two sub-processes: winner determination and payment mechanism. For winner determination, the auctioneer selects some bidders as winners and allocates items (money) to them for exchange. For payment

P. Lin et al., *Auction Design for the Wireless Spectrum Market*, SpringerBriefs in Computer Science, DOI 10.1007/978-3-319-06799-5_2, © The Author(s) 2014

mechanism, the auctioneer decides how much a winning buyer should pay for an item and how much a seller should get for a sold item.

2.2 Definitions in Auction Mechanisms

Definition 2.1. Collusion: A ring that some bidders form to bid against outsiders to gain extra benefits by manipulating the auction result.

In the research literature, it is assumed that there is no "Collusion" of bidders. Each bidder plays for himself, which make the auction look like a non-cooperative game.

Definition 2.2. Valuation: Bidders each have a value in mind which represents how much an item is worth to them. The value is their evaluation of the item.

Usually, the value is private information. Some works assume that the value follows some distribution and the distribution is known by the auctioneer. Therefore a buyer is not willing to pay more than a predetermined value to get an item. Likewise, a seller is not willing to sell the item at a lower than the predetermined price (called "reserve price"). This is a required economic property and should be satisfied in the design.

Definition 2.3. Individual Rationality: an auction has individual rational if no buyer pays more than his bid and no winning seller is paid less than his ask.

The Individual Rationality property guarantees that if bidders and sellers submit their true value, they will not receive negative utility, which provides them some incentives to act truthfully. We define the utilities as follows.

Definition 2.4. Utility: for a bidder, the utility is the difference between the true value of all the winning items and the total payment; for a seller, we define the utility as the total income from sales (some work may define it as the difference between the total income and the true value of all the sold items); for the auctioneer, we define the utility as the difference between the total payment from the buyers and the total income of the sellers.

There is an economic property in relation to the auctioneer's utility.

Definition 2.5. Ex-post Budget Balance: a double auction is ex-post budget balanced if the auctioneer's utility can never be negative.

This property ensures the auctioneer has an incentive to set up the auction. In practice the auctioneer can charge a transaction fee. In the research literature, people adopt "Ex-post Budget Balance" for simplicity.

Let us introduce a concept used in both game theory as well as in auction.

Definition 2.6. Dominant Strategy: a dominant strategy is the one that maximize a players utility regardless of what other players' strategies are. Mathematically, if x_i is player i's strategy, for any $x_i' \neq x_i$, and any strategy profile of others x_{-i}, we have $U_i(x_i, x_{-i}) \geq U_i(x_i', x_{-i})$.

Based on "Dominant Strategy", we define "Truthfulness":

Definition 2.7. Truthfulness: an auction is truthful if the true value is the players dominant strategy.

When designing auction mechanisms, it is crucial to make them truthful. As the most important property, it has been well accepted in the research literature [21, 23] for the following reasons. As the auctioneer does not know the buyers' private values, truthfulness is the best method to prevent market manipulation, which can hurt the interests of other buyers and sellers. Besides, truthfulness simplifies the strategic decision process for all players, as their true values are the best strategies.

The "Efficiency" property represents the performance of an auction.

Definition 2.8. Efficiency: an auction is efficient if the aggregate of all participants' utilities are maximized.

It shares a common meaning with "Social Welfare" which describes the aggregate utility of all buyers, sellers and the auctioneer. As the payment among them can be canceled out, "Efficiency" depends only on the determination process of who is the winner. If the items are allocated to those who value them most, then the auction is efficient.

We regard "Efficiency" as the second most important property as we always trade-off between truthfulness and efficiency in auction design. To achieve truthfulness, we compromise efficiency. Usually it is not difficult to design a truthful auction, but it is very challenging to prove that the auction is the most efficient one among all the truthful mechanisms. Very few works have paid attention to making the efficiency as high as possible even though they ensure the truthfulness. Therefore truthful auctions are often designed with possibly very low efficiency.

Besides "Efficiency", "Time Complexity" is also an easily ignored property in the research literature.

Definition 2.9. Time Complexity: the time complexity of an (auction) algorithm quantifies the amount of it takes to run as a function of size of the input to the problem. An algorithm with low time complexity is computationally tractable.

In practice, we should consider the application scenario of the auction mechanism. If an auction is held frequently or its input size is too large, the mechanism itself should be computationally tractable.

2.3 Research Literature

Traditional auctions can be classified by the number of participants (buyers and sellers), and the number of commodities and their properties (i.e., homogeneous or heterogeneous, super-additive or sub-additive). Here we would like to focus on how auction mechanisms are introduced to the spectrum market and the necessary improvement of rules to accommodate the spectrum trading.

Most early works discussed single-seller multi-buyer auctions with homogeneous channels. Zhou et al. [21] and Jia et al. [10] are two representative works. Zhou et al. [21] proposed VERITAS to support an eBay-like dynamic spectrum market. VERITAS is truthful mechanism and has a polynomial complexity of $O(n^3k)$ with n bidders and k channels. Different from FCC-style spectrum auction which is for the long-term and large geographical regions, it aims for the short-term and small regions. The auction complexity should be low enough to obtain a result quickly. The interference relationship is represented by a conflict graph. Most of the latter works in this area have followed their dead with short-term small-region auctions, and using a conflict graph as pre-knowledge. In [10] a VCG-like auction is used to maximize the expected revenue of the seller given the distribution of the buyers' evaluations. Considering that VCG auctions are usually computationally intractable, they further designed a truthful suboptimal auction with polynomial time complexity. Zhu et al. in [23] extended the meaning of a buyer from a single node to a multi-hop secondary network. They designed a truthful auction for trading homogeneous channels between a seller and multiple secondary networks.

Later, double auction mechanisms (multi-seller multi-buyer) have also been considered in the spectrum market. Zhou et al. [22] proposed TRUST for spectrum trading. It satisfies good properties such as spectrum reusability, truthfulness, budget balance and individual rationality, but it sacrificed one group of buyers (taking its bid as the clearing price) to achieve truthfulness. We believe it is the first to propose the grouping method to tackle the interference relationship. Any two users that might interfere with each other should not be placed in the same group. The conflict graph is then divided into several independent sub-graphs. Although it indeed achieves truthfulness, efficiency is sacrificed, because they can use only random grouping, and the grouping itself affects the efficiency of the auction. For example, who should be grouped together and how many groups should there be? Following the idea of TRUST, Wu et al. [16] improve up on it by only sacrificing one buyer in each group, and at the same time achieving truthfulness. Both works inherited the McAfee mechanism [12], which required homogeneity of channels. Yang et al. [18] were the first to consider the auction with heterogeneous channels. They proposed TASC, a mechanism extending the McAfee mechanism, for a cooperative communication scenario. However, it restricted a unique clearing price for all the channels, which could seriously reduce the system efficiency when the budget and evaluations of the channels varied too much over a big range. Feng et al. [8] and Chen et al. [4] extended this work by considering spectrum reusability and diversity of channel characteristics.

Auction mechanisms are also integrated into the mechanism design in many research directions of networks. For example, Huang et al. [9] proposed a strategy-proof and privacy preserving auction for spectrum trading. Zhuo [24] and Dong [7] used an auction-based framework to motivate third-parties to lease their unused resources to service providers for dynamic cellular offloading. Li [11] and Wang [15] studied auction mechanism in cloud computing scenarios. In [11], cloud providers can buy/sell their Virtual Machines' capabilities on the open market as well as job scheduling and resource pricing. In [15], the authors made optimal capacity segmentation in cloud platform. Resources are priced in the pay-as-you-go style, or auctioned on the spot market, or by subscription, such that users' various types of demands are well satisfied. Wang [14] and Zhang [19] proposed auction in online fashion to satisfy continuously arriving demands.

2.4 Our Starting Points

Spectrum is a special commodity in that two devices using the same spectrum frequency can interfere each other if they are in close proximity. That also means if the devices are far away from each other, they can reuse the spectrum. This is the major difference between spectrum and other commodities. Researchers have extensively exploited the "interference" property in their works [4, 21]. The most frequently used model is Interference Graph, by which they make spectrum allocation and avoid interference as well as encourage spectrum reuse. The "reusability" property is also well utilized [16, 22]. The general method is to randomly allocates buyers into some independent groups such that buyers in the same group do not interfere with each other. The buyers in the same group will win the same channel or lose in the auction.

Our works focus on exploiting other features of spectrum auction, that is the heterogeneous propagation property, the group-buying model and the dynamic spectrum demand property. We do not consider either interference or reusability in the second and the third work for simplicity. The new properties themselves build complete stories.

2.4.1 Auction for Heterogeneous Spectrums

There are many existing works relating to our work that is presented in Chap. 3. However, most of them fail to consider spectrums as non-identical items. In [22], Zhou et al. first address spectrum reusability in their auction design: TRUST. In [17], the authors also consider spectrum reusability for buyers, and they assume buyers can have multiple radios. Recently, in [6], Dong et al. address spectrum reusability in a time-frequency division manner and model the problem as a combinatorial auction. The proposed TAHES scheme also considers spectrum reusability, moreover, TAHES can tackle the case where spectrums are heterogeneous.

In [13], an auction design for heterogeneous TV white space spectrums is proposed. In the paper, the spectrum allocation problem has been defined as an optimization problem where maximum payoff of the central trading entity (called spectrum broker) is the optimization goal. However, [13] is not a double auction scheme and its design goal is different from TAHES. Recently, in [18], Yang et al. proposed a double auction design for cooperative communication with heterogeneous relay selections. However, there is no reusability in their scenario.

Different from our single-round auction model, there have also been works considering spectrum auction in an online fashion [5, 14]. In an online spectrum auction, buyers may arrive at different times and request the spectrum for a particular duration. However, existing online double auction schemes consider only the homogeneous spectrum.

2.4.2 Spectrum Group-Buying Framework

We consider a spectrum group-buying scenario in secondary networks. The auction involves multiple small buyers, one seller, and multiple heterogeneous channels. Single buyers cannot afford a whole channel, nor would they need to use it exclusively. Thus a group-buying model is necessary to enable the function.

From all the state-of-the-art auction mechanism designs, we can see that most VCG-based auctions guarantee truthfulness, and maximize social welfare. However, they have some limitations, such as high computational complexity (exponential running time), and low revenue for the sellers (especially when there are less demands from buyers). Most recent works [10, 23] do not apply the original VCG auctions. Instead, VCG-like heuristic auctions are designed to approximate VCG auction at lower computational overhead. VCG double auction does not satisfy the budget balanced property especially when the demands of the buyers are low. The McAfee mechanism and its successors can work well in auctions with homogeneous items, but cannot be directly applied to those with heterogeneous items.

In summary, none of the existing works provide a feasible tool or solution to tackle the group-buying problem that we are discussing. Consequently, we design new algorithms SAMU and DCP for the two stages, which improve system efficiency significantly.

2.4.3 Flexible Auction

We find that most of previous auction works make the assumption that buyers can claim at most one channel. Therefore these mechanisms cannot satisfy the flexible demands of buyers. To the best of our knowledge, the only mechanism that enables a flexible demand is combinatorial auction. In this model, the authors provide time-frequency flexibility for buyers [6], where each buyer can submit a bid to claim time-channel combinations. The authors in [20] also proposed a simplified combinatorial auction model to make the multi-channel trading without a uniform price for each channel. The major concern regarding combinatorial auction is that in general cases, the problem is NP hard, which means that the mechanism is not suitable for periodic auctions with too many buyers and channels.

In all these works, an assumption is made that the buyers themselves will consume the channels for data transmission, therefore their satisfaction levels are quantitatively analyzed and considered in their utility functions. Usually, the satisfaction level is an input into the auction problem, known as the true evaluation.

However, in our scenario the operators provide services to users to make revenue. Naturally there are no true evaluations for channels at the very beginning and in fact they depend on their users' willingness to pay. We theoretically derive the operators' true evaluations of channels by analyzing the service provision, that is, the pricing mechanism and users' utility maximization.

In summary, none of the existing works provided a feasible auction model to enable flexible demand with polynomial time complexity. Most auction models are polynomial time efficient but they do not support flexible demand. Combinatorial auction supports flexible demand, but they are not computationally efficient. Flexauc preserves good properties of both types and provides flexibility with polynomial time (more specifically, linear time) algorithm. We also show how should the operators determine the best bids in the auction in the three-layered market.

2.5 Special Characters of Spectrum Auction

2.5.1 *Interference and Reusability*

We can liken interference and reusability to the two sides of a coin. For any pair of nodes in a network, we can set a threshold to distinguish interfering pairs from non-interfering pairs. Then the non-interfering pairs can reuse the spectrum frequency.[1]

Recall that "interference can occur when two sources in close proximity transmit simultaneously at the same frequency". So to avoid interference, the transmissions can be isolated by spatial domain, temporal domain, frequency domain, or code domain. In the research literature of spectrum auction, researchers focus on spatial and temporal reusability, which do not need to involve extensive knowledge of technical aspects, while the other two require a deep understanding of the basis of wireless communication and signal processing.

An auction itself can often be formulated as an optimization problem [6]. The objective function is selected as social welfare or the strong side's utility. The "interference and reusability" properties are usually formulated as constraints. Thus the auctioneer needs to solve the optimization problem(s) to obtain the winners and their payments. Here we focus on the constraints of the optimization problem. Usually they are based on either spatial or temporal reusability.

1. **Spatial reuse**: is the most frequently used as it enables the non-interfering nodes that transmit at the same channel simultaneously. Usually researchers assume a pre-knowledge of conflict graph, which describes the interference relationship of the nodes. We represent that two nodes can interfere with each other by an edge between them in the graph. If there is no edge between them, they can reuse the same frequency band or channel.

[1]Note that it is a simplified analytical method which are frequently used in studying resource allocation of wireless networks.

If there is only one channel, the auctioneer prefers to allocate it to the non-interfering nodes making total bids most attractive. The allocation problem turns to the maximum weighted independent set problem, which is an NP-hard optimization problem.

If there are multiple channels to trade, the frequently used method is to divide the nodes into several independent sets (groups). Each group can obtain some channel(s) and a random matching is used to map the groups and channels to make the auction truthful. If the matching depends on the bids, the mechanism usually leaves room for untruthful bidding. Typical works using this method include [16, 22].

The conflict graph can also be used as constraints for the optimization problem in resource allocation. Let a_{ij} denote the allocation state of node i on channel j. If $a_{ij} = 1$, then i node gets j channel. Otherwise, it cannot use j channel.

Obviously the sufficient and necessary conditions that the optimization problem should satisfy are: for any edge ik in the conflict graph and any j,

$$a_{ij} + a_{kj} \leq 1 \tag{2.1}$$

holds. The constraints can be further simplified: for any clique C in the graph,

$$\sum_{k \in C} a_{kj} \leq 1 \tag{2.2}$$

holds. "Clique" means the sub-graph which is also a complete graph. Equation (2.2) is a sufficient and necessary condition to guarantee the spectrum allocation is effective and interference-free. Here the term "sufficient" means that any array $\{a_{ij}\}$ that satisfies Eq. (2.2) is an interference-free allocation. The term "necessary" means that any interference-free allocation can be represented by an array $\{a_{ij}\}$ that satisfies Eq. (2.2).

2. **Temporal reuse**: it is relatively less used. Temporal reuse means the interfering nodes transmit in the same channel at a different time. The input of this problem can be the whole period of time (assumed to be [0, 1]) and channel states. The output of this problem is the channel allocation result in the temporal domain, for example, node i_1 uses channel j_1 in the period of $[0.3, 0.5]$. As no interfering pairs can simultaneously use the same channel, we use the variable $a_{ij} \in [0, 1]$ to denote the time portion of the whole period when channel j is allocated to node i.

   However, in this case, Eq. (2.2) is necessary but not sufficient. Figure 2.1 a counterexample showing that an array satisfies Eq. (2.2) but does not make itself an interference-free allocation. We know that $\{a_{ij}\} = \{0.4, 0.6, 0.4, 0.6, 0.4\}$ satisfies Eq. (2.2). However, when making allocation, node 1 occupies the channel in the period of $[0, 0.4]$, node 2 in $[0.4, 1]$, node 3 in $[0, 0.4]$, node 4 in $[0.4, 1]$, and node 5 in $[0, 0.4]$. We find that node 1 and node 5 interfere with each other.

Fig. 2.1 A counterexample of Eq. (2.2)

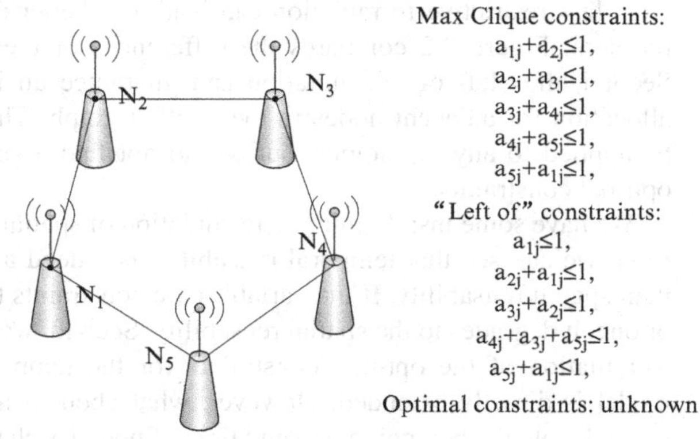

Fig. 2.2 An illustration of the constraints

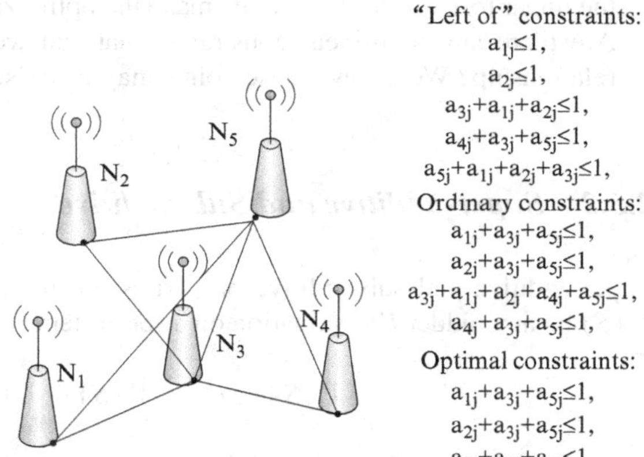

An alternative method to model the temporal reuse, named "left of" constraint, is sub-optimal [2, 3]. The "left of" constraints are

$$a_{ij} + \sum_{k \in C_i} a_{kj} \leq 1, \forall i, j \tag{2.3}$$

where C_i is the set of node i's "left of" and interfering neighbors. Node k is in C_i if and only if node k is to the left of node i (in a geographic location) and the two nodes form an interfering pair. For ease of understanding, "left of" formulation imposes a topological order among nodes. The advantages of "left of" constraints are three-fold. First, it is less restrictive than ordinary constraint formulation, though it may not be optimal.[2]

[2]With ordinary constraints' formulation, each node involves any interfering BSs into its constraints. So it is more restrictive than "left of" ones. The optimal constraints are the least restrictive ones which also guarantee an interference-free allocation. These constraints are shown in Fig. 2.2.

A less restrictive formulation can lead to a better result of the optimization problem. Figure 2.2 compares the differences of the constraint formulations. Second, the "left of" formulation can guarantee an interference-free channel allocation for adjacent nodes in the conflict graph. Third, this formulation can be applied to any topologies, but we do not find a general formulation of the optimal constraints.

We have some insight into the formulation of spatial and temporal reusability. First, we can see that temporal reusability is indeed a more generalized model than spatial reusability. If the variables are constraints to the values of only zero or one, it degrades to the spatial reusability. Second, we do not find a generalized formulation of the optimal constraints for the temporal reusability when the model is directly extended. However, what about using two variables b_{ij} and e_{ij} to denote the beginning and end time of node i's channel allocation? Besides, the linear form of the constraints make the optimization problems easier to solve. Are there any non-linear constraints that can well formulate the interference relationship? We believe these points may give rise to future works.

2.5.2 Super-additive and Sub-additive

Super-additive and sub-additive properties are used in a multi-item auction. Let $V_i(S)$ be any bidder i's evaluation on a set of items S. For any extra item j, if

$$V_i(S \cup \{j\}) \leq V_i(S) + V_i(\{j\}) \tag{2.4}$$

always holds, then the super-additive property is satisfied. This property can be explained by the marginally decreasing effect in economics. When the items are substitutes, this property usually holds. If

$$V_i(S \cup \{j\}) \geq V_i(S) + V_i(\{j\}) \tag{2.5}$$

always holds, we say that the sub-additive property is satisfied. When the items are complements, for example, a set of stamps, this property usually holds.

These two properties cannot be emphasized in one design. They apply in the wireless network area as well as other areas, but as far as we know, relatively less works in spectrum auction exploit these properties. The authors in [6] proposed a combinatorial auction with time-frequency flexibility. Although they do not explicitly mention the super-additive property, their algorithm indeed implies it. In general, combinatorial auctions come with exponential computational complexity. In [6], they assume that bidders can only bid for continuous channels, such that the complexity is reduced to polynomial time. The physical meaning of this assumption is that communication technologies impose minimum bandwidth requirements on frequency bands. More continuous channels provide larger bandwidth, which facilitates service deployment, therefore the super-additive property holds. In our work "Flexauc" which is presented later, we study the sub-additive property in spectrum auction. The property is built on the basis of the elastic demand of the end users and the marginally decreasing effect. Similarly, with the help of this property and the assumption of identical items, we design a linear-time algorithm for the combinatorial auction.

2.5.3 Group Structure

As channels can be shared by non-interfering bidders, researchers usually define bidders who share the same channel as a group.

Note that the group structure can be exogenous or endogenous. The term "exogenous" means that the bidders have been divided into some groups and the group structure are the input of the auction problem. For example, in our work "group-buying framework" which is presented later, users belonging to the same secondary network are form within the same group and the auction takes this as input. The term "endogenous" means that the group structure is formed in the process of the auction. For example, in [16, 22], the auctioneer randomly divides non-interfering bidders into groups.

As far as we know, we are the first to study the exogenous group structure in spectrum auction. Similar to super-additive and sub-additive properties, it is not unique to wireless networks, so our mechanism can be applied to other scenarios with appropriate modification.

References

1. Auction. http://en.wikipedia.org/wiki/Auction.
2. C. Buragohain, S. Suri, C.D. Toth, and Y. Zhou. Improved throughput bounds for interference-aware routing in wireless networks. *Lecture Notes in Computer Science*, 4598:210, 2007.
3. L. Cao and H. Zheng. Distributed spectrum allocation via local bargaining. In *IEEE SECON*, pages 475–486, 2005.
4. Y. Chen, J. Zhang, K. Wu, and Q. Zhang. Tames: A truthful auction mechanism for heterogeneous spectrum allocation. In *INFOCOM*. IEEE, 2013.
5. Lara Deek, Xia Zhou, Kevin Almeroth, and Haitao Zheng. To preempt or not: Tackling bid and time-based cheating in online spectrum auctions. In *INFOCOM, 2011 Proceedings IEEE*, pages 2219–2227. IEEE, 2011.
6. M. Dong, G. Sun, X. Wang, and Q. Zhang. Combinatorial auction with time-frequency flexibility in cognitive radio networks. In *INFOCOM, 2012 Proceedings IEEE*, pages 2282–2290. IEEE, 2012.
7. W. Dong, S. Rallapalli, R. Jana, L. Qiu, K. Ramakrishnan, L. Razoumov, Y. Zhang, and T. Cho. Ideal: Incentivized dynamic cellular offloading via auctions. In *INFOCOM*. IEEE, 2013.
8. X. Feng, Y. Chen, J. Zhang, Q. Zhang, and B. Li. Tahes: A truthful double auction mechanism for. heterogeneous spectrums. *IEEE Transactions on Wireless Communications*, 2012.
9. Q. Huang, Y. Tao, and F. Wu. Spring: A strategy-proof and privacy preserving spectrum auction mechanism. In *INFOCOM*. IEEE, 2013.

10. J. Jia, Q. Zhang, Q. Zhang, and M. Liu. Revenue generation for truthful spectrum auction in dynamic spectrum access. In *Proceedings of the tenth ACM international symposium on Mobile ad hoc networking and computing*, pages 3–12. ACM, 2009.
11. H. Li, C. Wu, Z. Li, and F. Lau. Profit-maximizing virtual machine trading in a federation of selfish clouds. In *INFOCOM*. IEEE, 2013.
12. R.P. McAfee. A dominant strategy double auction. *Journal of economic Theory*, 56(2): 434–450, 1992.
13. Marcin Parzy and Hanna Bogucka. Non-identical objects auction for spectrum sharing in tv white spaces – the perspective of service providers as secondary users. In *New Frontiers in Dynamic Spectrum Access Networks (DySPAN), 2011 IEEE Symposium on*, pages 389–398. IEEE, 2011.
14. S Wang, P Xu, X Xu, S Tang, X Li, and X Liu. Toda: truthful online double auction for spectrum allocation in wireless networks. In *New Frontiers in Dynamic Spectrum, 2010 IEEE Symposium on*, pages 1–10. IEEE, 2010.
15. W. Wang, B. Li, and B. Liang. Towards optimal capacity segmentation with hybrid cloud pricing. *University of Toronto, Tech. Rep*, 2011.
16. F. Wu and N. Vaidya. A strategy-proof radio spectrum auction mechanism in noncooperative wireless networks. *IEEE Transactions on Mobile Computing*, 2012.
17. Fan Wu and Nitin Vaidya. Small: A strategy-proof mechanism for radio spectrum allocation. In *INFOCOM, 2011 Proceedings IEEE*, pages 81–85. IEEE, 2011.
18. D. Yang, X. Fang, and G. Xue. Truthful auction for cooperative communications. In *Proceedings of the Twelfth ACM International Symposium on Mobile Ad Hoc Networking and Computing*, page 9. ACM, 2011.
19. H. Zhang, B. Li, H. Jiang, F. Liu, A. Vasilakos, and J. Liu. A framework for truthful online auctions in cloud computing with heterogeneous user demands. In *INFOCOM*. IEEE, 2013.
20. T. Zhang, F. Wu, and C. Qiao. Special: A strategy-proof and efficient multi-channel auction mechanism for wireless networks. In *INFOCOM*. IEEE, 2013.
21. X. Zhou, S. Gandhi, S. Suri, and H. Zheng. ebay in the sky: strategy-proof wireless spectrum auctions. In *Proceedings of the 14th ACM international conference on Mobile computing and networking*, pages 2–13. ACM, 2008.
22. X. Zhou and H. Zheng. Trust: A general framework for truthful double spectrum auctions. In *INFOCOM 2009, IEEE*, pages 999–1007. IEEE, 2009.
23. Y. Zhu, B. Li, and Z. Li. Truthful spectrum auction design for secondary networks. In *INFOCOM*. IEEE, 2012.
24. X Zhuo, W Gao, G Cao, and Y Dai. Win-coupon: An incentive framework for 3g traffic offloading. In *Network Protocols (ICNP), 2011 19th IEEE International Conference on*, pages 206–215. IEEE, 2011.

Chapter 3
Truthful Double Auction Mechanism for Heterogeneous Spectrums

In this chapter, we consider the problem of redistribution of spectrums via a double auction and propose truthful auctions mechanism for TV channels.

It is a promising way to enable spectrum owners to lease their spectrums to secondary service providers where there are no interference to the primary users. In return, the spectrum owners can get paid from secondary service providers. An example is the Spectrum Bridge [2] company. It has launched an online platform to allow spectrum owners to lease unused spectrums to buyers. Bands from different sellers are in different frequencies and with various availabilities. Also, there are different preferences from spectrum buyers for different bands. Moreover, the spectrums can be reused. Two buyers that are not conflict with each other may use the same band at the same time.

We can model the spectrum redistribution between spectrum owners and buyers as a *single round multi-item double auction*, where the spectrum owners are the sellers; the secondary service providers are the buyers; the spectrums are the goods.

Although auction has been widely studied, we can not directly apply existing double auction schemes [3, 8, 15, 17, 19, 20] to the new scenario. We face three major challenges here. First, we need to deal with the spatial heterogeneity. The available bands to buyers can be different due to different owners. However, in [3, 8, 15, 17, 19, 20] only the scenario where all spectrums are available to all buyers is considered. Second, there is frequency heterogeneity and spectrum reusability. Due to the reason that the spectrums are from various frequency bands and with different communication ranges (low frequency band have larger transmission rages). Therefore, in different bands, the interference relationships between different buyers can be different. However, existing works usually assume the same conflict range for all frequencies. Third, auction mechanism design under this scenario is challenge. The most critical property: *Truthfulness* (or strategy-proofness) should be preserved. A truthful auction incites all bidders to voluntarily reveal their true valuation for the items they are bidding. Unfortunately, the truthfulness is hard to achieve applying existing schemes when there are heterogeneous items for sale. Besides truthfulness, several other properties are also desirable: (i) *Individual rationality*: The utility of both buyers and sellers is enhanced because of the auction;

P. Lin et al., *Auction Design for the Wireless Spectrum Market*, SpringerBriefs in Computer Science, DOI 10.1007/978-3-319-06799-5_3, © The Author(s) 2014

(ii) *Budget balance*: The net profit of the auctioneer is non-negative; and (iii) *System Efficiency*: the total utility of all buyers and sellers is optimized.

In this chapter, we propose a **T**ruthful double **A**uction scheme for **HE**terogeneous **S**pectrum called TAHES to solve the above challenges. In TAHES spectrum buyers are grouped according to their conflict relationships in heterogeneous bands to explore spectrum reusability. We also introduce a matching procedure between buyer groups and sellers to guarantee truthfulness. Furthermore, there is a novel pricing scheme integrated in TAHES to improve the system efficiency.

We organize the rest of the chapter as follows. In Sect. 3.1 we present the auction model and then set the design objectives. Section 3.2 further explains challenges in heterogeneous auction design. Detailed description of TAHES is given in Sect. 3.3. Simulation results are provided in Sect. 3.4. Section 3.5 summarizes the entire work.

3.1 Model Description and Design Targets

In this section, The problem of heterogeneous spectrum exchange between spectrum owners and service providers is formulated as a double auction.

3.1.1 Problem Formulation

In the scenario, there are M spectrum owners selling spectrum bands to N buyers. The seller and buyers participate in a single-round double auction for this purpose. We use $S = \{s_1, s_2, \cdots, s_M\}$ to denote the set of sellers and $Z = \{z_1, z_2, \cdots, z_N\}$ to denote the set of buyers. There is also an auctioneer to decide the winning bidders and the payment. Here, we simply assume that each seller has one channel to sell and each buyer wants to buy one channel and The auction is sealed-bid, private and collusion-free. In another word, in the auction, all bidders simultaneously submit sealed bids so that no bidder knows any other participants' bids. In addition, we assume that bidders do not collude with each other to improve the utility of the coalitional group. There are different valuations from the buyers for the channels.

In the auction, the auctioneer determines the payment P_i^s for seller s_i and the price p_i^z that buyer z_i should pay. P_i^s and p_i^z are not necessarily equal. We use b_i^j to denote the bid of z_i for seller s_j's channel. $B_i = (b_i^1, b_i^2, \cdots, b_i^M)$ is the bid vector of z_i. $B = (B_1, B_2, \cdots, B_N)$ is the bid matrix of all buyers. Let B_{-i} denote the bid matrix with buyer z_i's bid B_i excluded. We use C_i to denote the bid of s_i. $C = (C_1, C_2, \cdots, C_M)$ is the bid matrix of all sellers and C_{-i} denotes the bid matrix with s_i's bid removed. The true valuation of s_i for its channel is V_i^s and the true valuation of z_i for seller s_j's channel is v_i^j. $V_i^z = (v_i^1, v_i^2, \cdots, v_i^M)$ is the valuation vector of buyer z_i. If s_j's channel is unavailable to z_i, v_i^j is zero. The true valuation of both buyers and sellers may or may not be equal to their bids.

Therefore, we define the utility of buyer z_i as:

$$u_i^z = \begin{cases} v_i^{\theta(i)} - p_i^w, & \text{if } z_i \text{ wins} \\ 0, & \text{Otherwise} \end{cases} \tag{3.1}$$

in which $\theta(i)$ is the channel that z_i wins. Similarly, the utility of the seller s_i is:

$$U_i^s = \begin{cases} P_i^s - V_i^s, & \text{if } s_i \text{ wins} \\ 0, & \text{Otherwise} \end{cases} \tag{3.2}$$

3.1.2 Design Target

Truthfulness, budget balance and system efficiency cannot be achieved in any double auction at the same time, even without the consideration of individual rationality [11, 18]. Since we need to warrant that spectrum owners have the incentive to lease their spectrum and also the third party (spectrum transaction platform, government) is willing to participate as an auctioneer in our scenario, in this chapter, our design target are to achieve truthfulness, individual rationality and budget balance:

- *Truthfulness*. Neither sellers nor buyers can get higher utility by not reporting their true valuation, i.e., $C_j \neq V_j^s$ or $B_i \neq V_i^z$.
- *Individual rationality*. Winning sellers and winning buyers obtain non-negative utility from the auction.
- *Budget balance*. There is non-negative profit for the auctioneer, which equals to the price paid by the buyers minus the payment to the sellers.

3.2 Challenges of Heterogeneous Spectrum Auction Design

In this section, the challenges of designing a truthful auction mechanism for heterogeneous spectrum are briefly illustrated.

3.2.1 Spatial Heterogeneity

The meaning of spatial heterogeneity is that the spectrum availability is different in different locations. For example, one TV channel is available only if there are no nearby TV stations or wireless microphones using the same channel.

In traditional auction design, spectrum reusability is usually treated by grouping buyers together leveraging independent sets on the interference graph (or conflict

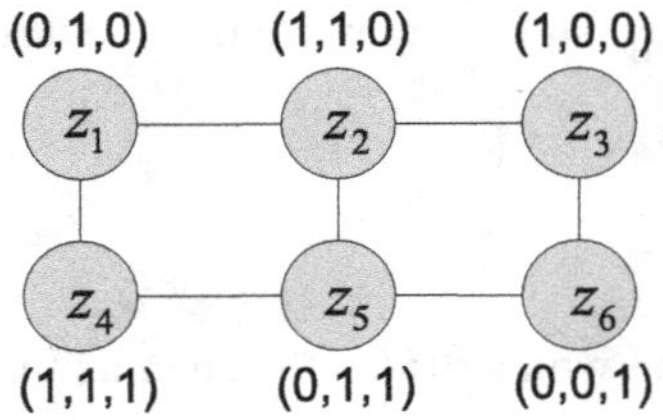

Fig. 3.1 There are no channels commonly available for all buyers in different groups: $\{z_1, z_3, z_5\}$ or $\{z_2, z_4, z_6\}$

graphs) [13], and then the group bid is set according to the minimum bid from the buyers in the group [16, 20]. However, if there no common available channels for buyers in a group, the group bid should be 0. As a result, this group can never win the auction. See the example in Fig. 3.1. There are three channels: 1–3 and 6 buyers: $z_1 - z_6$. The bid vectors are also depicted along with each buyer. If the two nodes are mutually conflict, there will be an edge between them, e.g. only channel 2 is available for buyer z_1, so only its bid for channel 2 is not 0.

In this example, the six buyers can be grouped into two groups, $\{z_1, z_3, z_5\}$ and $\{z_2, z_4, z_6\}$, by finding the maximal independent sets in the interference graph. However, there is no common available channel for all users in either group. For example, in the group $\{z_1, z_3, z_5\}$, channel 2 is the only common available channel for the two users z_1 and z_5, but it is not available for z_3.

3.2.2 *Frequency Heterogeneity*

The meaning of frequency heterogeneity is that there are different transmission ranges for different frequencies. Based on the model recommended by ITU [5], the center frequency of one spectrum band can impact the path loss between two nodes:

$$L = 10 \log f^2 + \gamma \log d + P_f(n) - 28 \tag{3.3}$$

where γ is the distance power loss coefficient, $P_f(n)$ is the floor loss penetration factor, f is the frequency of transmission in megahertz(MHz), L is the total path loss in decibel(dB), and d is the distance in meter(m). In our model, the spectrums sold by different owners are from a wide range of frequencies. For example, in the German spectrum auction held in 2010, the highest frequency (2.6 GHz) was more than three times higher than the lowest frequency (800 MHz) [1]. There will be a more than 10 dB path loss difference at the same distance between 800 MHz and 2.6 GHz frequencies, which will lead to non-identical interference relationships among spectrum buyers on the two channels.

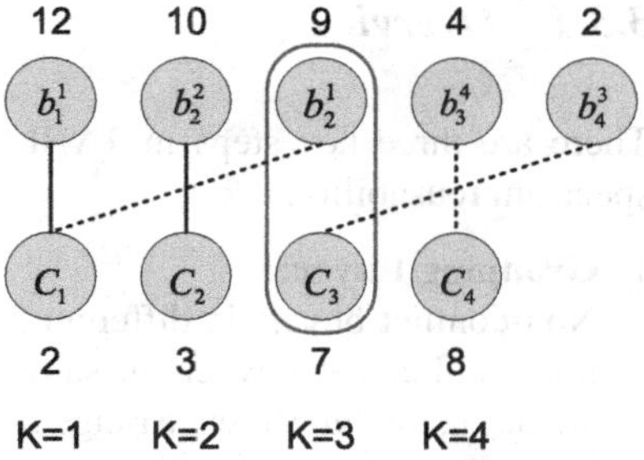

Fig. 3.2 Buyer b_2 can increase utility by manipulating its non-winning bid

3.2.3 Market Manipulation

In a double auction for homogeneous items, truthfulness can be ensured by the two well-known auction algorithms VCG [4,7,14] and McAfee [10]. However, they may not be truthful if we directly apply these schemes to heterogeneous items.

In the McAfee double auction, the auctioneer sorts the buyers' bids in non-increasing order and the sellers' bids in non-decreasing order: $b_{i_1}^{j_1} \geq b_{i_2}^{j_2} \geq \cdots \geq b_{i_{N\times M}}^{j_{N\times M}}$ and $C_{i_1} \leq C_{i_2} \leq \cdots \leq C_{i_M}$. Then the auctioneer finds the largest k such that $b_{i_k}^{j_k} \geq C_{j_k}$. The McAfee scheme discards $(b_{i_k}^{j_k}, C_{j_k})$ which is the least profitable transaction and sets the uniform price for buyers to be $b_{i_k}^{j_k}$ and the uniform payment for sellers to be C_{j_k}. In a double auction for heterogeneous items, although one buyer can win at most one item, it can manipulate its bids to achieve higher utility.

In Fig. 3.2, there are four buyers among which buyer z_2 has two non-zero bids b_2^2 and b_2^1. Both the buyers and sellers' bids have been sorted according to the McAfee mechanism. We can see that the index of the least profitable transaction marked by the red frame is $k = 3$. So the price charged for the winning group is $b_2^1 = 9$. If the buyer z_2 lowers its bid for seller s_1 to make $b_2^1 < 9$, say $\tilde{b}_2^1 = 8$, this misconduct will not change the results of the auction, but can contort the price from 9 to 8. As a result, the utility of z_2 can be increased.

The VCG double auction suffers the same defect. Moreover, is scheme also fails to achieve the budget balance [18,20].

3.3 Auction Design: TAHES

In this section, we introduce the design of TAHES.

3.3.1 *Overview*

There are three key steps in TAHES to handle both spectrum heterogeneity and spectrum reusability:

1. **Grouping Buyer:**
   Non-conflict buyers in different locations can reuse spectrums. The meaning of non-conflict is that when the same channel h is used by two buyers z_i and z_j, they are not in the interference range of each other. However, the conflict relationships between one pair of buyers are non-identical in different frequencies. In this step, a grouping algorithm is used by the auctioneer considering non-identical interference graphs to form non-conflict buyer groups. So that the buyers in the same group can purchase the same channel. The grouping algorithm should bid-independent. The input is the channel availability information of each buyer, which can be calculated according to the path loss model given the locations of buyers and sellers.
2. **Matching Buyers and Sellers:**
   After the first step, if the buyers in a group have more than one common channels each buyer group may still purchase channels from multiple sellers. However each group can at most win one channel. To solve this issue, in this matching step, the auctioneer chooses one conventional matching algorithm to match each buyer group to only one seller. The matching should be made based only on the channel availability for each group. In another word, the matching step is also bid-independent.
3. **Winner Determination and Pricing:**
   After the first two steps, the remaining problem of winner determination and pricing will be similar to that in the McAfee double auction. However, since each buyer group in our problem is matched to one seller, if we directly use the kth pair of buyer group and seller to determine the winners, the number of winning bidding pairs may be small. So our design further enhances the McAfee mechanism by taking into consideration of the matching results.

3.3.2 *Auction Procedure*

TAHES comprises the following steps:

3.3.2.1 Buyer Grouping

Suppose the set of channels from sellers is $H = \{h_1, h_2, \cdots, h_M\}$. h_i's communication range which is defined as the transmission range under the maximum allowed power level on channel h_i is denoted as $R(h_i)$. Without loss of generality, we assume $R(h_1) \leq R(h_2) \leq \cdots \leq R(h_M)$. Let $A = \{a_{i,j} | a_{i,j} \in \{0, 1\}\}_{N \times M}$, an N by

Algorithm 1 Buyer-Grouping(A, E, H)

// L represents the set of grouped buyers
$L = \emptyset, G = \emptyset, F = \emptyset$;
while $L \neq Z$ **do**
  for all $h_i \in H$ **do**
    Candidate buyers to be grouped: $Q = \emptyset$;
    $Q = \{z_k | A_{k,i} = 1 \wedge z_k \notin L\}$;
    Find independent set IS_i on buyer set Q based on E_i;
  end for
  Find IS_i, such that $\mathbb{EFF}(IS_i)$ is maximized;
  $g = \{z_j | z_j \in IS_i\}$;
  $L = L \bigcup \{z_j | z_j \in IS_i\}$;
  $f = \{h_j | h_j \in H \wedge R(h_j) <= R(h_i)\}$;
  $G = G \bigcup g, F = F \bigcup f$;
end while
return (G, F);

M matrix, represent the buyers' channel availability. $a_{i,j} = 1$ means that channel h_j is available for buyer z_i. Let $E = \{e_{i,j,k} | e_{i,j,k} \in \{0, 1\}\}_{M \times N \times N}$, an M by N by N matrix, represent the conflict relationships between buyers in each channel. $e_{i,j,k} = 1$ means that buyers z_j and z_k are conflict in h_i.

In this step, the inputs are A and E, which are both bid-independent. After grouping, we get a set of l ($l \leq N$) buyer groups denoted as $G = \{g_1, g_2, \cdots, g_l\}$ and the corresponding candidate channel set for each group denoted as $F = \{f_1, f_2, \cdots, f_l\}$. f_i contains the channels that g_i can purchase, which is assigned by the auctioneer. G and F are the outputs of the grouping algorithm. The grouping algorithm should satisfy the following constraints:

Interference Free Constraint: Any two buyers in the same group do not mutually interfering with each other in any channel in the candidate channel set.

$$\forall z_j, z_k \in g_i, \forall h_l \in f_i \wedge A_{j,l} = A_{k,l} = 1 \Rightarrow E_{l,j,k} = 0 \tag{3.4}$$

Common Channel Existence Constraint: There exists at least one channel that is available for all buyers in the same group.

$$\forall g_i, \exists h_j, s.t. \forall z_k \in g_i \Rightarrow h_j \in f_i \wedge A_{k,j} = 1 \tag{3.5}$$

The grouping algorithm first finds an independent set of buyers in each channel. Then it selects one such set with maximum *Grouping Efficiency* and continues to find the next group until all buyers are classified into one group. The efficiency of each group relates to the channel communication range and group size. If the independent set found on channel h_i is IS_i, the efficiency can be defined as:

$$\mathbb{EFF}(IS_i) = |IS_i| \times R(h_i) - \alpha \cdot V(G, h_i) \tag{3.6}$$

Algorithm 2 Buyer-Group-Matching(G, F, S)

1: // Let Δ be an $|G|$ by $|S|$ matrix representing the weighted adjacent matrix between G and S
2: $\Delta = \{0\}_{M \times N}$;
3: **for all** $g_x \in G, s_y \in S$ **do**
4: **if** $\delta_x^y > 0$ and $h_y \in f_x$ **then** $\Delta_{x,y} = \Delta_{y,x} = |g_x|$;
5: **end for**
6: $(G_C, S_C, \sigma) = MATCH(X, Y, \Delta)$;
7: return (G_C, S_C, σ);

where $V(G, h_i)$ is the number of groups already formed according to the independent set on the interference graph of h_i. From this definition, we can see that it is preferred to group buyers on a channel with a larger communication range. This is because if IS_i is an independent set of the conflict graph of channel h_i, it is also an independent set of the conflict graph of any channel $h_j (j < i)$. It is also preferred to group buyers into various channels since one channel can only be sold to one buyer group. We can set the parameter α to be very large, e.g. $N \times R(h_M)$, to ensure an even distribution of groups on all channels.

The grouping procedure is shown in Algorithm 1.

In Algorithm 1, we actually can use any existing algorithms to find independent sets (e.g., the algorithms in [12]). From the procedure of Algorithm 1, it is obvious that all buyers in IS_i have a common available channel h_i. And from line 12, we can see that the grouping algorithm only considers the candidate channels with a smaller or equal communication range of h_i. Therefore, the grouping results G and F also satisfy the Interference Free Constraint.

Theorem 3.1. *The buyer groups and candidate channel sets returned by Algorithm 1 satisfy the Common Channel Existence Constraint and the Interference Free Constraint.* □

3.3.2.2 Matching

After step 1, buyers form a group set G. If we assume the number of buyers in group g_i is n_i and the group bid vector is $\delta_i = (\delta_i^1, \delta_i^2, \cdots, \delta_i^M)$. The group bid is assigned to be the minimum bid times the group size as:

$$\delta_i^j = \min\{b_k^j | z_k \in g_i\} \cdot n_i \tag{3.7}$$

Here, each buyer can be viewed as one *super buyer*. In δ_i, there may be more than one non-zero entry. If not well designed, the auction may be untruthful, as shown in Sect. 3.2.3. Such a type of market manipulation is caused by the multiple non-zero group bids. To tackle this challenge, TAHES leverages a channel matching scheme. It matches one buyer group to an identical seller. The results after matching the candidate winning group are referred to as set G_C and candidate winning sellers are referred to as set S_C. We show the matching procedure in Algorithm 2.

Algorithm 3 Winner-Determination-and-Pricing(G_C, S_C, σ)

1: $G_Z = \emptyset$, $S_Z = \emptyset$; // The set of winning groups and sellers
2: Construct $X = \{g_{i_1}, g_{i_2}, \cdots, g_{i_l}\}$, such that $\delta_{i_1}^{\sigma(i_1)} \geq \delta_{i_2}^{\sigma(i_2)} \geq \cdots \geq \delta_{i_l}^{\sigma(i_l)}$;
3: Construct $Y = \{s_{j_1}, s_{j_2}, \cdots, s_{j_M}\}$, such that $C_{j_1} \leq C_{j_2} \leq \cdots \leq C_{j_M}$;
4: Find the largest k, s.t. $\delta_{i_k}^{\sigma(i_k)} \geq C_{j_k}$;
5: **if** $k < 2$ **then** return $(G_Z, S_Z, 0, 0)$;
6: Find the groups Λ_k^Z, s.t. $\forall g_x \in \Lambda_k^Z$, $\delta_{i_k}^{\sigma(i_k)} = \delta_x^{\sigma(x)}$;
7: Find the sellers Λ_k^S, s.t.$\forall s_y \in \Lambda_k^S$, $C_{j_k} = C_y$;
8: // Let X_i be the sublist of the first i groups in X, Y_j be the sublist of the first j sellers in Y
9: Find X_k, Y_k, s.t. $|M(X_{k-1}, Y_{k-1})|$ is maximal, where $g_i \in \Lambda_k^Z$ and $s_j \in \Lambda_k^S$ can be in any orders;
10: // Determine the winning groups and sellers
11: $p^z = \delta_{i_k}^{\sigma(i_k)}$, $P^s = C_{j_k}$, $G_Z = \{g_x | \forall g_x \in M(X_{k-1}, Y_{k-1})\}$, $S_Z = \{s_y | \forall s \in M(X_{k-1}, Y_{k-1})\}$;
12: // Determine the price and payment
13: **for all** Buyer $z_i \in Z$ **do**
14:     **if** $g_{\tau(i)} \in G_Z$ **then** $p_i^z = p^z / n_{\tau(i)}$;
15: **end for**
16: **for all** Seller $s_j \in S$ **do**
17:     **if** $s_j \in S_Z$ **then** $P_j^s = P^s$;
18: **end for**
19: return (G_Z, S_Z, P^S, p^z);

In this algorithm, σ records the matching result. For example, $\sigma(g_x) = s_y$ indicates that buyer group g_x is assigned to seller s_y. $MATCH(X, Y, \Delta)$ matches nodes set X to Y with weighted edges in Δ. It can be any matching algorithm for bipartite graphs specified by the auctioneer, for example, maximum matching [6] or maximum weighted matching [9]. This matching step here is also independent both the buyers and sellers' bids.

3.3.2.3 Winner Determination and Pricing

In the winner-determination and pricing stage, we need to consider the matching results from the second step. Here, we can directly apply the pricing mechanism used in McAfee here. However, the winner determination algorithm in McAfee may reduce the number of winning pairs and thus decrease the system efficiency. Observed that there may be multiple equal g_{i_k}, s_{j_k} pairs, the auction outcome may be affected by the order of bids after sorting. So we check all the possible orders of bids equal to g_{i_k} or s_{j_k} and choose the one with the maximum matchings. The detailed algorithm is shown in Algorithm 3.

In line 9, the algorithm checks all the combinations of possible orders of groups in Λ_k^Z and sellers in Λ_k^S to determine the number of matchings that can be achieved. The algorithm finally finds the maximum number of matchings for the given winner candidate sets G_C, S_C. $M(X, Y)$ denotes the set of matching induced by X and Y. In line 6, the algorithm finds all group bids equal to $\delta_{i_k}^{\sigma(i_k)}$, and stores them in the set

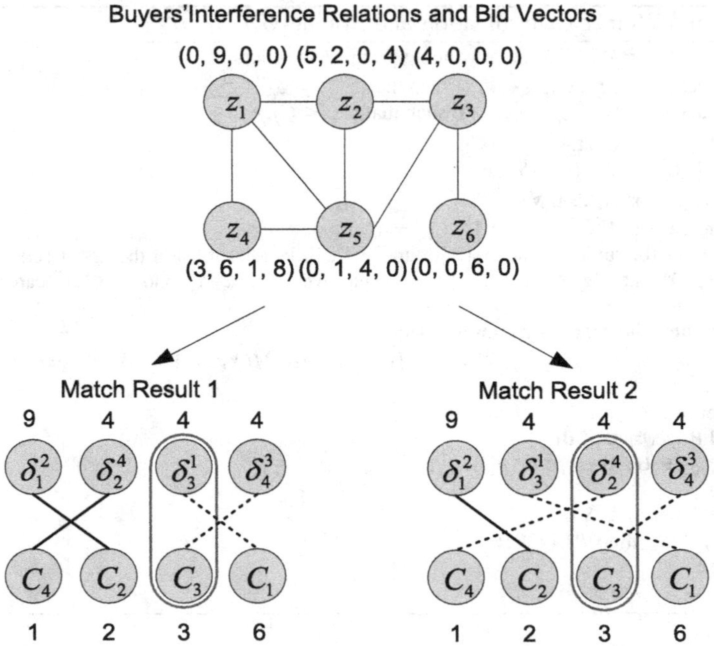

Fig. 3.3 An illustrative example with six buyers

Λ_k^Z. Actually, the order of group bids in Λ_k^Z can be arbitrary. Similarly, Λ_k^S contains all sellers with bids equal to C_k. From lines 13 to 18, the algorithm calculates the price for each buyer and the payment for each seller. $\tau(\cdot)$ is the mapping function from the indices of buyers to the indices of the groups they belong to. The group price is shared by all the members in that group.

3.3.3 Illustrative Example

Figure 3.3 shows a scenario with six buyers and four sellers. The bid vectors and buyers' conflict relationships are shown in Fig. 3.3. For the sake of simplicity, here we assume their conflict relationships are the same across all channels. The sellers' bids are: $s_1 = 6$, $s_2 = 2$, $s_3 = 3$ and $s_4 = 1$.

After buyer grouping, the buyers form four groups with group bids shown in Table 3.1. The bids in bold font are chosen after the matching step. We can see that, there are three candidate bids with the same value 4. In the winner-determination and pricing step, after sorting all the bids, according to Algorithm 3, in our scenario, $k = 3$, $\Lambda_k^Z = \{g_2, g_3, g_4\}$, $\Lambda_k^S = \{s_3\}$. Then the algorithm checks all possible orders of groups in Λ_k^Z and orders of sellers in Λ_k^S to determine the winners.

Table 3.1 Members and bids of four buyer groups

Groups	s_1	s_2	s_3	s_4
$g_1 = \{z_1\}$	0	**9**	0	0
$g_2 = \{z_2, z_4\}$	3	2	0	**4**
$g_3 = \{z_3\}$	**4**	0	0	0
$g_4 = \{z_5, z_6\}$	0	0	**4**	0

When g_2 orders first in Λ_k^Z, there are two successful matchings, as shown in Matching Result 1. Otherwise, there can only be one successful matching. One of such cases is shown in Matching Result 2. Therefore, we will use the order in Matching Result 1 as the auction result. The price for one winning group $p^z = \delta_3^1 = 4$, the payment for the winning seller $P^s = C_3 = 3$. The price for winners g_2 and g_4 will be shared by the members in these two groups. Therefore, $p_2^z = p_4^z = p_5^z = p_6^z = 2$, $p_1^z = p_3^z = 4$. The profit the auctioneer makes is $2 \times (4 - 3) = 2$.

3.3.4 Proofs of Economic Properties

In this section, we prove that TAHES is individually rational, budget-balanced and truthful.

Theorem 3.2. *TAHES is individually rational.* □

Proof. For each buyer z_i in winning group g_k, we have: For each winning seller s_j: $P_j^s \geq C_k \geq P^s$

$$p_i^z = \frac{p^z}{n_k} \leq \frac{\delta_k^{\sigma(k)}}{n_k} \leq \frac{n_k \cdot b_i^{\sigma(k)}}{n_k} = b_i^{\sigma(k)}$$

Theorem 3.3. *TAHES is budget-balanced.* □

Proof. According to the sorting in the winner determination algorithm, we have $p^z \geq P^s$ and $|G_Z| = |S_Z|$, therefore the budget for the auctioneer is: $|G_Z| \times p^z - |S_Z| \times P^s \geq 0$

Theorem 3.4. *TAHES is truthful.* □

To prove Theorem 3.4, we first show that the auction result for buyer z_i is only related to the bid $z_i^{\theta(i)}$ ($\theta(i)$ is the channel that z_i wins, $\theta(i) = \sigma(\tau(i))$). Then we show that the winner determination is monotonic and the pricing is bid-independent, such that for any buyer z_i or seller s_j, it cannot increase its utility by bidding untruthfully.

Lemma 3.1. *The position of k stays the same for all possible combinations of orders of Λ_k^Z and Λ_k^S.*

Proof. Since the group bids of groups in set Λ_k^Z are equal and the seller bids of sellers in set Λ_k^S are also equal. The order of elements in Λ_k^Z and Λ_k^S do not change k.

Lemma 3.2. *After matching between buyers and channels, the auction result only depends on the buyers' bids for the assigned channel.* □

Proof. Both the grouping and matching step are bid-independent. After grouping, buyer z_i is in group $g_{\tau(i)}$. After the matching step, only the group bid $g_{\tau(i)}^{\theta(i)}$ is considered in the winning determination and pricing procedure. $g_{\tau(i)}^{\theta(i)}$ is only related with buyer bid $s_i^{\theta(i)}$.

Lemma 3.3. *Given B_{-i}, if buyer z_i wins in the auction, it also wins by bidding $\widetilde{b_i}^{\theta(i)} > b_i^{\theta(i)}$.* □

Proof. Case 1: if $b_i^{\theta(i)} = g_{\tau(i)}^{\theta(i)}$, $g_{\tau(i)}^{\theta(i)}$ can only be increased for $\widetilde{b_i}^{\theta(i)}$. Since the channel matching is independent of bids, the buyer group $g_{\tau(i)}$ will also be matched to the same seller. As the group bid increased, during the winning determination stage, the buyer group $g_{\tau(i)}$ matching to $s_{\theta(i)}$ can still wins the auction. Case 2: if $b_i^{\theta(i)} > g_{\tau(i)}^{\theta(i)}$, $\widetilde{b_i}^{\theta(i)}$ will not change $g_{\tau(i)}^{\theta(i)}$.

Lemma 3.4. *Given C_{-j}, if seller s_j wins in the auction, it also wins by bidding $\tilde{C}_j < C_j$.* □

Proof. Since the channel matching is independent of bids, the seller s_j should be matched to the same buyer group when bidding lower. Since its bid is decreased, during the winning determination stage, the buyer group $g_{\sigma^{-1}(j)}$ matching to s_j can still win.

Lemma 3.5. *Given B_{-i}, if buyer z_i wins the auction by bidding $\widetilde{b_i}^{\theta(i)}$ and $b_i^{\theta(i)}$, the prices charged to z_i are the same.* □

Proof. According to Lemmas 3.1 and 3.3, increasing a winning buyer's bid will not change the auction results. Therefore k will not change. Since the price is only dependent on k, the determined prices will also be the same.

Lemma 3.6. *Given C_{-j}, if buyer s_j wins the auction by bidding $\tilde{C}_j$ and C_j, the payment paid to s_j are the same.* □

Proof. According to Lemmas 3.1 and 3.4, decreasing a winning seller's bid will not change the auction results or k. Similarly with Lemma 3.5, the prices charged are the same.

Lemma 3.7. *TAHES is truthful for buyers.* □

Proof. We need to prove that no buyer b_i can increase its utility by bidding untruthfully, that is, when bidding $\widetilde{b_i}^{\theta(i)} \neq b_i^{\theta(i)}$, $\tilde{u}_i^z < u_i^z$.

Case 1: $\tilde{u}_i^z = 0$, $u_i^z > 0$.
Case 2: $\tilde{u}_i^z = u_i^z = 0$.

Case 3: According to Lemma 3.5, $\tilde{u}_i^z = u_i^z > 0$

Case 4: According to Lemma 3.3, it happens only when $\tilde{b}_i^s > v_i^{\theta(i)}$. Since z_i wins the auction by bidding higher, z_i should have offered the lowest bid in the group $g_{\tau(i)}$ when bidding truthfully. As a result, $\delta_{\tau(i)}^{\theta(i)} = n_{\tau(i)} \cdot v_i^{\theta(i)}$. Since z_i wins when bidding untruthfully and loses when bidding truthfully, its group bid should satisfy the following condition: $\tilde{\delta}_{\tau(i)}^{\theta(i)} \geq \tilde{\delta}_x^{\sigma(x)} \geq \delta_{\tau(i)}^{\theta(i)}$. Here we suppose (x, y) are the boundary in the auction result. The price paid by z_i when bidding untruthfully is $p_i^z = \tilde{\delta}_x^{\sigma(x)} / n_{\tau(i)} \geq \delta_{\tau(i)}^{\theta(i)} / n_{\tau(i)} = v_i^{\theta(i)}$. Therefore, $\tilde{u}_i^z < 0 = u_i^z$.

Given Lemma 3.2, a buyer cannot improve its utility by submitting any bid vector other than its true valuation vector.

Similarly, we can prove the following lemma on the truthfulness for sellers.

Lemma 3.8. *TAHES is truthful for sellers.* □

Proof of Theorem 3.4: Lemmas 3.7 and 3.8 together prove that TAHES is truthful.

3.4 Numerical Results

In this section, simulations are designed to evaluate the performance of TAHES. The truthfulness of TAHES will first be verified. Then we study the economic impact on the system efficiency.

3.4.1 Simulation Settings

In the simulation, buyers are randomly distributed in a 400×400 square. There are 5–30 spectrum sellers. In most cases, the radio interference range spans from 20 to 100. Each seller has its own base stations in the same area, therefore, the channel is only available to buyers when there is no primary base station nearby. The number of base stations of one seller can be from 2 to 5.

We assume, in the auction, the buyers' bids are randomly distributed over $[0, V_{MAX})$ where V_{MAX} can be within the range of [4, 10]. If the channel is not available for the buyer, the bid is 0. The sellers' bids are randomly distributed over $[2, 2 \times V_{MAX})$. Due to spectrum reusability, sellers may expect its channel to be sold to more than one buyers, so we set the sellers' bids higher. We set $V_{MAX} = 6$ as a default. We use Maximum Matching [6] as the default in both Algorithms 2 and 3. In our simulation, all the results are averaged over more than 100 runs.

The metrics evaluated in the simulation are as follows:

- Buyer satisfaction ratio: percentage of buyers that can get one channel.
- Number of successful transactions.
- Average group size.

Table 3.2 TAHES is truthful for buyers

M	$U_T^w > U_U^w$ (%)	$U_T^w = U_U^w$ (%)	$U_T^w < U_U^w$ (%)
5	0.9	99.1	0
10	0.4	99.6	0
15	0.4	99.6	0

Table 3.3 TAHES is truthful for sellers

M	$U_T^w > U_U^w$ (%)	$U_T^w = U_U^w$ (%)	$U_T^w < U_U^w$ (%)
5	7.3	92.7	0
10	11.2	88.8	0
15	10.6	89.4	0

The first two metrics together determine the system performance of the auction. The third one reflects the effectiveness of the grouping algorithm.

3.4.2 *Truthfulness of TAHES*

To show the truthfulness of TAHES, we fix the number of buyers as 50. The number of sellers can be 5, 10 and 15. In each run, we randomly sample three buyers and three sellers and compare their utilities when bidding truthfully or untruthfully.

The results for the buyers are shown in Table 3.2, in which U_T^w (U_U^w) means the utility when a buyer is bidding truthfully(untruthfully). We observe (i) the utility of one buyer when bidding truthfully is always higher than that when bidding untruthful, and (ii) the utilities of one buyer when bidding truthfully or untruthfully are the same for most of the time. The first observation verifies the truthfulness of TAHES for buyers. The reason for the second phenomenon is that one buyer can hardly change the group utility by changing its bid.

The results for sellers are shown in Table 3.3, in which U_T^s (U_U^s) means the utility when a seller is bidding truthfully(untruthfully). It shows that TAHES is also truthful for sellers.

3.4.3 *Impact on System Performance*

In this section, we evaluate the system performance of TAHES based on two metrics: buyer satisfaction ratio and number of successful transactions.

3.4.3.1 Impact of the Winner Determination Algorithm

The winner determination algorithm used in TAHES finds an optimal order to determine winners. However, the McAfee mechanism just consider one possible order of bids. In the simulation, we vary the buyer number N and fix $M = 10$.

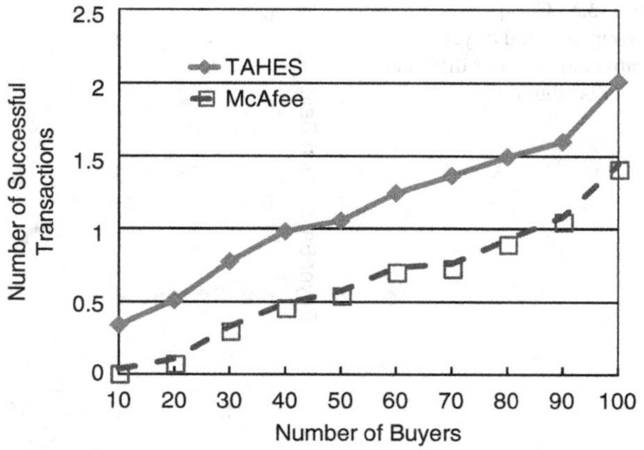

Fig. 3.4 Comparison between TAHES and McAfee

Figure 3.4 shows the number of successful transactions provided by both TAHES and McAfee. In all node densities, TAHES provides 30 % more transactions than McAfee. In this sense, our winner determination algorithm has traded the time complexity for the system efficiency.

$$\begin{cases} \mathbb{EFF}-1 = |IS_i| \times R(h_i) - \alpha \cdot V(G, h_i) \\ \mathbb{EFF}-2 = |IS_i| \times R(h_i) \\ \mathbb{EFF}-3 = |IS_i| - \alpha \cdot V(G, h_i) \\ \mathbb{EFF}-4 = |IS_i| \\ \mathbb{EFF}-5 = \begin{cases} \mathbb{EFF}-1, & \text{if } |IS_i| \leq N/M \\ 0, & \text{otherwise} \end{cases} \end{cases} \tag{3.8}$$

3.4.3.2 Impact of the Bid Distribution

In previous simulations, we fix V_{MAX} to be 6. Here we vary V_{MAX} from 4 to 10 while $N = 50$. Figure 3.5 illustrates the buyer satisfaction ratio in different V_{MAX}. With higher V_{MAX}, the expected value of buyer's valuation is higher. However, since the group bid is determined by the minimum bid in the group, whose expected value does not increase at the same speed as V_{MAX}. Therefore, we can observe that the buyer satisfaction ratio has decreased for larger V_{MAX}.

3.4.3.3 Impact of the Communication Range

To evaluate the impact of the communication range to the auction result, we change the span of channels' communication from [20–100] to [40–200]. We fix the seller number $M = 10$. Figure 3.6 shows the satisfaction ratios. With a larger

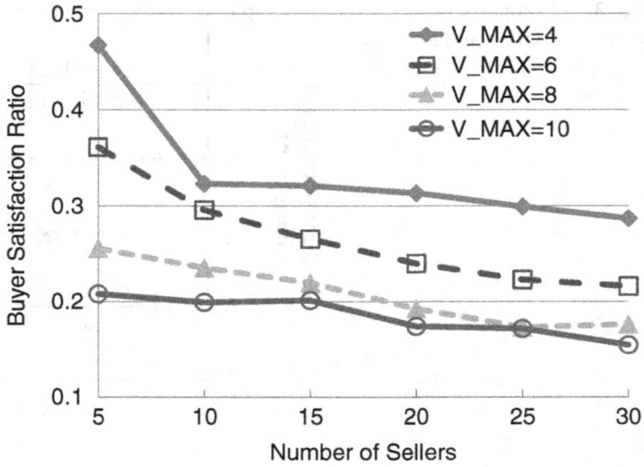

Fig. 3.5 Comparison of group size and buyer satisfaction under different bid distributions

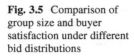

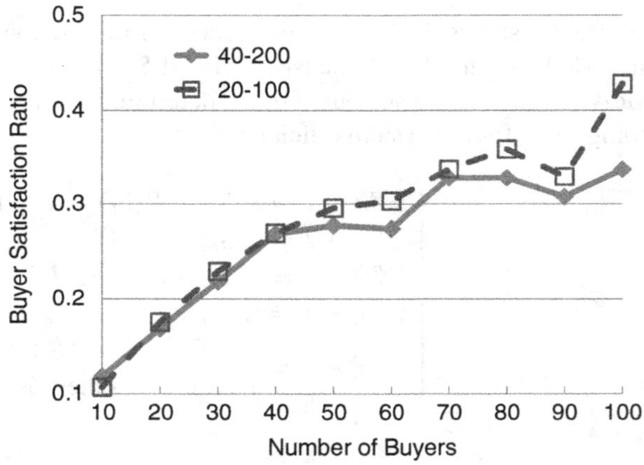

Fig. 3.6 The impact of communication range on buyer satisfaction

communication range, the degree of spatial reuse decreases, so we can observe that the span [20–100] provides a higher buyer satisfaction ratio compared with that of [40–200].

3.4.3.4 Impact of the Grouping Function $\mathbb{EFF}(\cdot)$

In the buyer grouping step, we have used the idea of *grouping efficiency* to determine which buyers are grouped on which channel first. Other than the one used in our algorithm, there may be some other choices shown in Eq. (3.8):

$\mathbb{EFF}-1$ is the one used as our default efficiency function. α is set to $N \times R(h_M)$. Among the five functions, $\mathbb{EFF}-1$, $\mathbb{EFF}-3$, $\mathbb{EFF}-5$ try to distribute buyer groups evenly to different sellers. $\mathbb{EFF}-1$ and $\mathbb{EFF}-2$ both consider group size and channel interference range and prefer to group buyers under the channel with the

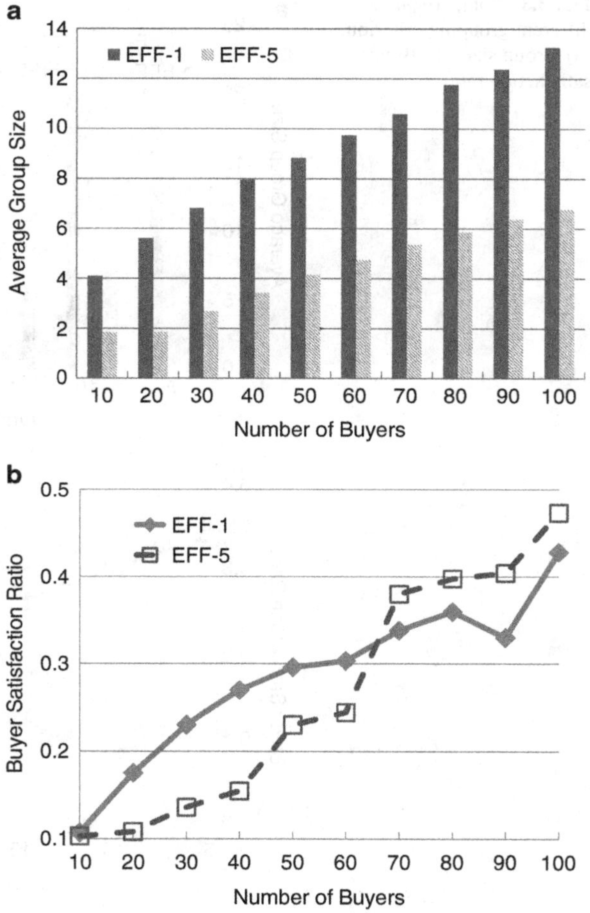

Fig. 3.7 Comparison between $\mathbb{EFF}-1$ and $\mathbb{EFF}-5$ in different node densities. (**a**) Group size. (**b**) Buyer satisfaction ratio

longer range. The first four functions are independent of seller number and they use maximum independent set of buyers as groups. On the other hand, $\mathbb{EFF}-5$ limits the size of a group to be no more than N/M. Therefore, the number of groups obtained from $\mathbb{EFF}-5$ is comparable with the number of sellers.

Figure 3.7 compares $\mathbb{EFF}-1$ and $\mathbb{EFF}-5$ in different node densities when $M = 10$. We can see from Fig. 3.7a, the group size by $\mathbb{EFF}-5$ is much smaller than that of $\mathbb{EFF}-1$ and is always below N/M. Interestingly, when the number of sellers is larger than 60, the efficiency provided by $\mathbb{EFF}-5$ outperforms that of $\mathbb{EFF}-1$. We guess there may be an optimal group size setting in different settings of M and N. We will further explore this issue in our future works.

Figure 3.8 shows the results under various M when $N = 50$. In Fig. 3.8a, $\mathbb{EFF}-1$ and $\mathbb{EFF}-2$ have smaller group size compared with $\mathbb{EFF}-3$ and $\mathbb{EFF}-4$. This is because the channel with the larger interference range may have

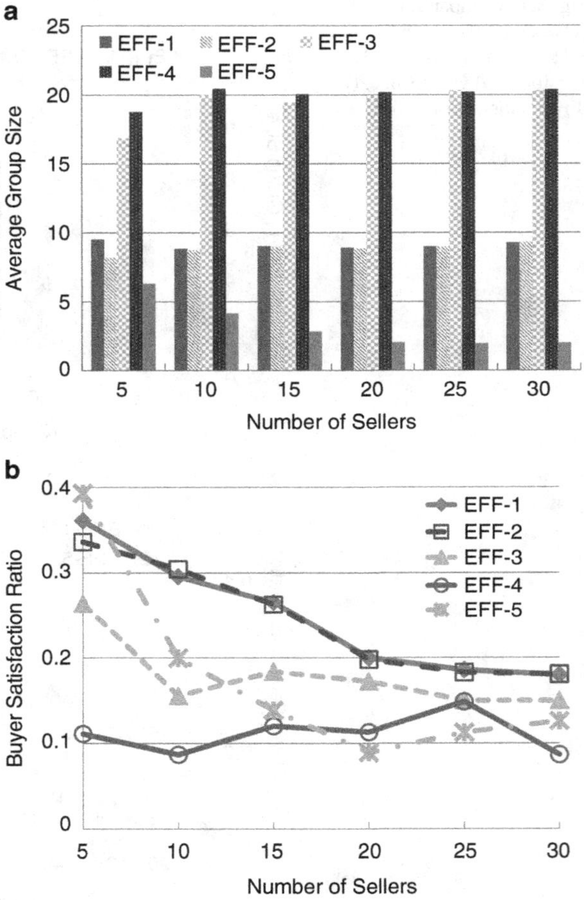

Fig. 3.8 Comparison of different grouping criterion. (**a**) Group size. (**b**) Buyer satisfaction ratio

more conflict neighbors, the group size of the channel with the larger interference range is expected to be smaller. As the group size increases, the total number of groups decreases, since the groups are matched to one dedicated seller, therefore, fewer groups can win and the buyer satisfaction ratio may decrease. However, the group size obtained from $\mathbb{EFF}-5$ is too small. Although more groups win, the total number of winning buyers is still low. Figure 3.8b shows this situation.

3.5 Conclusions

In this chapter, we have designed TAHES, a truthful double auction scheme for heterogeneous spectrum. TAHES allows multiple spectrum owners with available spectrums in different locations and different frequency bands to participate in the spectrum leasing to secondary service providers. TAHES increases spectrum utilization through spectrum reuse. Carefully designed TAHES can not only solve unique challenges caused by spectrum heterogeneity but also preserve nice economic properties: Truthfulness, Budget Balance and Individual Rationality. With mathematical analysis and extensive simulations, we have shown that TAHES can achieve all required properties and provides higher system efficiency compared with traditional double auction schemes.

References

1. German spectrum auction ends but prices low. http://www.rethinkwireless.com/2010/05/21/german-spectrum-auction-ends-prices-low.htm.
2. Spectrum Bridge. Specex: The online marketplace for spectrum.
3. Lin Chen, Stefano Iellamo, Marceau Coupechoux, and Philippe Godlewski. An auction framework for spectrum allocation with interference constraint in cognitive radio networks. In *INFOCOM, 2010 Proceedings IEEE*, pages 1–9. IEEE, 2010.
4. E.H. Clarke. Multipart pricing of public goods. *Public choice*, 11(1):17–33, 1971.
5. Propagation Data. Prediction methods for the planning of indoor radiocomm. systems and radio local area networks in the frequency range 900 mhz to 100 ghz. *Recommendation ITU-R*, pages 1238–1, 1999.
6. Jack Edmonds. Paths, trees, and flowers. *Canadian Journal of mathematics*, 17(3):449–467, 1965.
7. T. Groves. Incentives in teams. *Econometrica: Journal of the Econometric Society*, pages 617–631, 1973.
8. J. Jia, Q. Zhang, Q. Zhang, and M. Liu. Revenue generation for truthful spectrum auction in dynamic spectrum access. In *Proceedings of the tenth ACM international symposium on Mobile ad hoc networking and computing*, pages 3–12. ACM, 2009.
9. T Kameda and I Munro. A o(ve) algorithm for maximum matching of graphs. *Computing*, 12(1):91–98, 1974.
10. R.P. McAfee. A dominant strategy double auction. *Journal of economic Theory*, 56(2): 434–450, 1992.
11. R.B. Myerson and M.A. Satterthwaite. Efficient mechanisms for bilateral trading. *Journal of economic theory*, 29(2):265–281, 1983.
12. Shuichi Sakai, Mitsunori Togasaki, and Koichi Yamazaki. A note on greedy algorithms for the maximum weighted independent set problem. *Discrete Applied Mathematics*, 126(2):313–322, 2003.
13. Anand Prabhu Subramanian and Himanshu Gupta. Fast spectrum allocation in coordinated dynamic spectrum access based cellular networks. In *New Frontiers in Dynamic Spectrum Access Networks, 2007. DySPAN 2007. 2nd IEEE International Symposium on*, pages 320–330. IEEE, 2007.
14. William Vickrey. Counterspeculation, auctions, and competitive sealed tenders. *The Journal of finance*, 16(1):8–37, 1961.
15. F. Wu and N. Vaidya. A strategy-proof radio spectrum auction mechanism in noncooperative wireless networks. *IEEE Transactions on Mobile Computing*, 2012.
16. Fan Wu and Nitin Vaidya. Small: A strategy-proof mechanism for radio spectrum allocation. In *INFOCOM, 2011 Proceedings IEEE*, pages 81–85. IEEE, 2011.
17. Yongle Wu, Beibei Wang, KJ Ray Liu, and T Charles Clancy. A multi-winner cognitive spectrum auction framework with collusion-resistant mechanisms. In *New Frontiers in Dynamic Spectrum Access Networks, 2008. DySPAN 2008. 3rd IEEE Symposium on*, pages 1–9. IEEE, 2008.

18. D. Yang, X. Fang, and G. Xue. Truthful auction for cooperative communications. In *Proceedings of the Twelfth ACM International Symposium on Mobile Ad Hoc Networking and Computing*, page 9. ACM, 2011.
19. X. Zhou, S. Gandhi, S. Suri, and H. Zheng. ebay in the sky: strategy-proof wireless spectrum auctions. In *Proceedings of the 14th ACM international conference on Mobile computing and networking*, pages 2–13. ACM, 2008.
20. X. Zhou and H. Zheng. Trust: A general framework for truthful double spectrum auctions. In *INFOCOM 2009, IEEE*, pages 999–1007. IEEE, 2009.

Chapter 4
Spectrum Group-Buying Framework

Auction is a scheme widely used in spectrum redistributions. In spectrum auctions, spectrum holders offer channels for potential buyers to purchase. Buyers are mostly big companies with sufficient budgets. However, there may also be small companies with limited budges interested in spectrum auctions. These small companies can be referred to as secondary users (SUs), which usually do not benefit from the auction. Motivated by the recent group-buying behaviors in the Internet based service, we argue that SUs with small budgets can form a group and take part in the spectrum auction as a whole to increase their winning chances in the auction. These SUs within the same group then evenly share the cost and benefit of the won spectrum. There are no existing auction models that can be applied in this scenario. There are three unique challenges to enable this idea. First, how can a group leader select the winning SUs and charge them fairly and efficiently? Second, how to guarantee truthfulness of users' bids? Third, how to match the heterogeneous channels to groups when one group would like to buy at most one channel.

In this work, we propose, a new auction framework for spectrum group-buying referred to as TASG, to address the above challenges. TASG will enable group-buying behaviors among SUs. There are three stages in TASG. In the first stage, an algorithm is proposed to determine the group members and bids for the channels. In the second stage, the auction is conducted between the group leaders and the spectrum holder. We also propose a novel winner determination algorithm for this stage. In the third stage, the winning spectrums are further distributed by the group leaders to the SUs in the group. The leader also collect fees from the SUs. We are able to prove that TASG possesses the following good properties: as truthfulness, individual rationality, improved system efficiency, and computational tractability.

4.1 Group-Buying Behavior of Users with Limited Budget

The need of spectrum redistribution comes from the need of spectrum resources to deploy wireless services. In the current spectrum allocation plan, almost all

P. Lin et al., *Auction Design for the Wireless Spectrum Market*, SpringerBriefs in Computer Science, DOI 10.1007/978-3-319-06799-5_4, © The Author(s) 2014

spectrums have been allocated to licensed users by the regulators. To overcome the wireless spectrum crisis, some under-utilized spectrums are taken back by the regulators for redistribution, e.g., the under-utilized TV channels [6]. Spectrum auction is a commonly used measure for spectrum redistributions, especially under the dynamic spectrum management context. Spectrum regulators, such as the FCC (Federal Communications Commission), have held many auctions to distribute spectrum bands in primary market. However, due to the high prices appeared in such auctions, the newly freed spectrums can only be afforded by some largest mobile companies. For example, in the recent German spectrum auction [1] carried out in 2010, 41 spectrum blocks were sold at a total price of $5.5 billion to only 3 mobile operators. As a result, small wireless service providers cannot enjoy any benefits. Similar cases happen in secondary market where many secondary users belonging to different networks compete for unused spectrum from a spectrum holder. Individual secondary user with limited budget cannot afford a whole spectrum block by his own.

Inspired by the emerging group-buying services on the Internet, e.g., Groupon [2], we propose that buyers can voluntarily form groups and to acquire and share the whole spectrum band sold in the spectrum auctions together. In this chapter, the scenario with several Secondary Networks (SNs) is considered. We assume that there are many Secondary Users (SUs) who are willing to buy the unused channels from a Spectrum Holder (SH). Since a whole channel is not affordable by any individual SU, those SUs within the same SN are grouped together. As a group, they try to group-buy some channels. The SNs can be viewed as the groups of users taking part in the same deal in Groupon. We further assume there is a Secondary Access Point (SAP) in each SN acting as a group leader. An SAP can decide which SUs can join the group and collect money from the grouped SUs. SAPs also act as agents to take part in the spectrum auction held by the SH. Aggregating all the money from the selected SUs, now the SAPs have a chance to bid the spectrum in the competition with other groups and major players in the wireless industry.

Unfortunately, although spectrum auction has been studied for many years, there are several unique characteristics making this group-buying scenario challenging.

First, both the price and the spectrum are shared by the SUs within an SN. Therefore, the number of grouped SUs within an SN can affect the SUs' satisfaction, which together with payment decides SUs' utilities. The group-buying concept is different from the spectrum spatial reusability design in the existing works [11, 13]. In their scenarios, buyers who do not interfere with each other are randomly selected to form a group. Each buyer in the same group enjoys the whole channel and pays the same amount. That means individual buyer may afford a whole channel by itself and the number of group members does not affect one's utility directly. Besides, in the scenario we are discussing, different SUs may have different budgets (the maximum amount of money they can pay) and evaluations of the whole channel (the benefit if it obtains the whole channel). It is therefore challenging for the SAP to make fair, efficient and valid decisions about which SUs should be grouped together.

Second, a well designed auction scheme should preserve truthfulness (or strategy-proofness), which means that the auction should incite all bidders to voluntarily submit their true valuation for the items they are purchasing. In another word, the participators should not be willing to lie in the auction. The truthfulness should be enforced by the scheme design. That is to say, in a truthful auction, the ones that lie about their true valuation cannot get a higher revenue compared with that they do not lie. However, in our scenario, the budget and evaluation should be both taken into consideration when bidding the channel. Therefore, there is a more challenging two-dimensional truthfulness problem. Both the truthfulness of budget and evaluation should be preserved by proper auction mechanisms. Although there are existing works considering two-dimensional truthfulness. For example, [4] considered bid-based and time-based truthfulness. Wang et al. [10] designed $m + 1$-price auction and guaranteed the quantity and price truthfulness. However, in our scenario, the two dimensions are coupled together with a connection of the number of selected SUs thus it cannot be simply addressed with existing auction models.

Third, the SH may have multiple heterogeneous channels to sell, each with different channel characteristics. Here we assume that although the SAPs submit their bids for all the channels, they are willing to buy at most one channel. The heterogeneity of the channels brings troubles to our mechanism design. Although there are recent works considering auction design for heterogeneous items [5, 12], they cannot be applied in the group-buying scenario directly.

In this work, we propose TASG, a **T**hree-stage **A**uction framework for **S**pectrum **G**roup-buying, to address the above mentioned challenges. In TASG, the group-buying is conducted through three stages. In stage I, the SUs submit two-dimensional bids in terms of budgets and evaluations to the SAPs. Then the SAPs decide how much it can offer for each channel when auctioning spectrums from the SH, as well as the winner set consisting of the selected SUs. The decisions are made considering the SUs' budgets and evaluations by our algorithm SAMU (Scarifying At least M Users). In stage II, auction is conducted between the SH and the SAPs. We design a new winner determination algorithm called DCP (Diverse Clearing Price) to improve the system performance (i.e., the number of successful transactions). In stage III, obtaining the auction results in stage II, the winning SAPs further determine how to serve and charge the SUs. Actually, there are two auction procedures in TASG. The inner auction is between the SH (as seller) and the SAPs (as buyers). The outer auction happens between the SAPs (as sellers) and the SUs (as buyers). Through mathematically analysis, we prove truthfulness for both the auctions. Note that, in the outer auction, the truthfulness for the SUs is two-dimensional truthfulness.

Our main contributions in this work are summarized as follows.

1. We propose TASG to formulate and solve the challenges. The framework ensures truthfulness in the two auction procedures. Besides, we also make the framework individual rational, budget balance and computationally tractable.
2. We propose two new algorithms, SAMU and DCP for the two auction procedures respectively. Compared with existing auction models, SAMU and DCP can improve system efficiency significantly.
3. Our simulation results verify the economical properties of TASG and demonstrate the efficiency of the proposed SAMU and DCP.

4.2 System Model

In this section, we first introduce the scenario we are considering. Then we list the definition of the strategies, utility functions. Finally we present the design objective of the auction scheme.

4.2.1 Problem Formulation

In this chapter, we consider a static scenario with one spectrum holder (SH). The spectrum holder will be the seller of the channels. We assume there are C channels. The channels are heterogeneous in terms of bandwidth, frequencies, maximum allowed transmission powers, and Signal-to-Interference-Noise-Ratios (SINRs). Therefore, the channels may be valued differently by different buyers. Also, there are N infrastructure-based SNs in the scenario. The SAPs are part of the infrastructures, which are owned by all the SUs. There are N_j SUs in the j-th SN ($1 \leq j \leq N$). We denote the SH's reserved prices for the channel as $\{r_k\}$ ($k = 1, \cdots, C$), which means the winners in the auction should pay at least the reserved prices. We assume each SN would like to buy at most one channel.

TASG consists of two single-round multi-item auctions. One is carried out between the SAPs and the SH, in which the SAPs are the buyers and the SH is the seller. We refer it as the *inner auction*. The other one is conducted between the SUs and their SAP, in which the SUs are the buyers and the SAP is the seller. We refer it as the *outer auction*.

In the j-th SN, the SU S_i^j's true evaluation of the k-th channel is $\tilde{v}_i^j(k)$. Although $\tilde{v}_i^j(k)$ may be larger than the reserved price r_k, a single SU usually cannot afford a whole channel as its limited budget $\tilde{d}_i^j(k)$ is far lower than r_k.

The SUs in the same SN can group-buy a whole channel and share the cost. All the SUs in the j-th SN submit their bids for k-th channel ($k = 1, \cdots, C$) to the SAP for participating in the group-buying activity. A bid for k-th channel from SU S_i^j consists of two parameters: $d_i^j(k)$ and $v_i^j(k)$, which are his claimed values for evaluation and budget. After the grouping, if their SAP fails to obtain a channel, no transactions will happen and all these SUs' have zero utilities. If their SAP obtains k-th channel, there will be some SUs obtaining the channels and have non-negative utility. These users are called the winner set $\mathbb{W}_j$. For the other SUs (in the loser set $\mathbb{L}_j$), they get nothing and their utilities are still zero.

The relationship between the shared capacity and utility is modeled with a linear model. The utility of an SU S_i^j is defined as the difference between the value of its share of the channel $\frac{\tilde{v}_i^j(k)}{|\mathbb{W}_j|}$ and its payment $\dot{d}_i^j(k)$, where $|\mathbb{W}_j|$ is the cardinal number of $\mathbb{W}_j$:

$$U_i^j = \begin{cases} \frac{\tilde{v}_i^j(k)}{|\mathbb{W}_j|} - \dot{d}_i^j(k) \text{ if } S_i^j \in \mathbb{W}_j \text{ and } \tilde{d}_i^j(k) \geq \dot{d}_i^j(k) \\ 0 \qquad\qquad \text{otherwise} \end{cases} \tag{4.1}$$

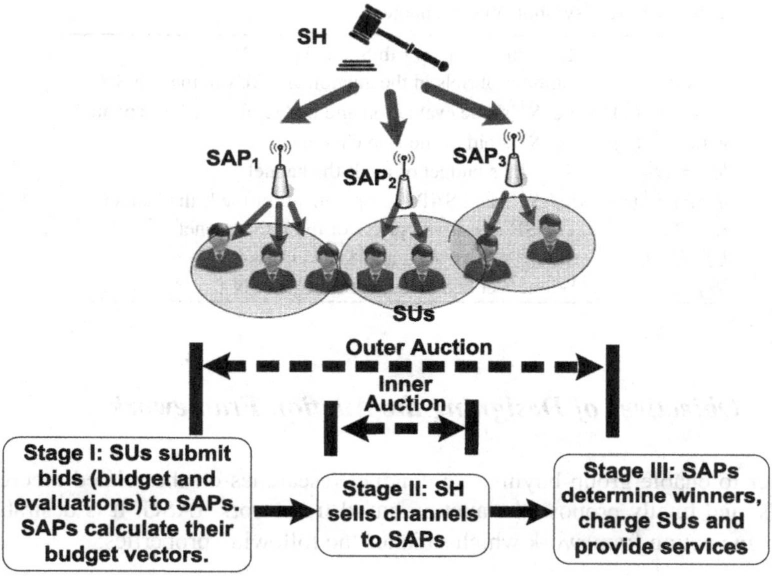

Fig. 4.1 A three-stage auction framework in TASG

$\tilde{d}_i^j(k) \geq \dot{d}_i^j(k)$ means that the winner should be able to pay the amount $\dot{d}_i^j(k)$. As the design of discriminatory pricing or sharing mechanism within the same SN usually does not satisfy the truthfulness requirement, we take the even sharing manner such that every winner gets the same share.

The SH acts as a seller and the auctioneer in the inner auction. We define the SH's final utility as $U = \sum_k U^k$, where U^k is its profit on the k-th channel.

$$U^k = \begin{cases} \dot{b}_j^k - r_k & \text{if sold to the j-th SAP} \\ 0 & \text{if not sold} \end{cases} \tag{4.2}$$

In the outer auctions, the SAPs function as sellers. They help to aggregate their SUs' bids. In the inner auction, the SAPs act as strategic buyers. The j-th SAP offers a bid vector $\mathbf{b}_j = \{b_j^k\}$, where b_j^k is its bid for the k-th channel. If it wins the k-th channel, the SAP gets a non-negative utility, which is defined as the difference between the aggregate amount it collects from SUs $\tilde{b}_i^k = \sum_{S_i^j \in \mathbb{W}_j} \dot{v}_i^j(k)$ and its payment $\dot{b}_j^k$ to the SH. Otherwise, the SAP's utility is zero. We define the j-th SAP's utility as (Table 4.1)

$$U_j = \begin{cases} \tilde{b}_j^k - \dot{b}_j^k & \text{if winning the k-th channel} \\ 0 & \text{otherwise} \end{cases} \tag{4.3}$$

Table 4.1 Key symbols in this chapter

S_i^j, S_j	The i-th SU in the j-th SN, the j-th SN
N, N_j	Number of SNs in the auction and SUs in the j-th SN
$\tilde{v}_i^j(k), \tilde{d}_i^j(k)$	SU S_i^j's true evaluation and budget for the k-th channel
$v_i^j(k), d_i^j(k)$	SU S_i^j's bid of the k-th channel
$\tilde{b}_i^j = H_k$	SAP P_j's budget of the k-th channel
$\dot{d}_i^j(k), \dot{b}_i^j(k)$	SU S_i^j's and SAP P_j's payment for the k-th channel
r_k	The SH's reserved price for the k-th channel
U_i^j, U_j, U	SU S_i^j's, SAP P_j's and SH's utilities
$\mathbb{W}_j, \mathbb{L}_j$	Winner set and loser set of the j-th SN

4.2.2 *Objectives of Designing the Auction Framework*

In order to enable group-buying, we conduct researches on the related theoretical models, and finally propose the auction based framework TASG. It is desirable to design an auction framework which satisfies the following properties:

1. Individual Rationality: Specifically, here it means $\tilde{v}_i^j(k) \geq \dot{v}_i^j(k)$, $\tilde{b}_j^k \geq \dot{b}_j^k$, and $\dot{b}_j^k \geq r_k$ in (4.1) and (4.2) respectively.
2. Truthfulness: Specifically, here we define the mechanism to be two-dimension truthful for SUs if any SU S_i^j's true values of channels $\{(\tilde{v}_i^j(k), \tilde{d}_i^j(k))\}$ is its dominant strategy.

Our primary objective is to satisfy as much SUs' demand for spectrum as possible and promote the market trading. We are originally motivated by the possibility that a whole channel from SH may sell at a high price, to the extent that a single SU cannot afford. It is promising to enable group-buying activities.

However, it is very challenging. As far as we know, all auctions that satisfy truthfulness sacrifice system efficiency more or less, such as VCG mechanism [3, 7] and McAfee mechanism [8]. It is even proved that it was impossible to achieve all the four properties for double auctions [9]. Following previous works, we believe that truthfulness should be placed in the top-priority position. Besides truthfulness, TASG satisfies individual rationality and budget balance, and it tries to improve the system efficiency and reduce time complexity under the constrains of the previous three properties.

4.3 Three-Layer Auction Framework

In this section, we propose TASG, a three-stage auction framework, In stage I, all SUs submit bids to their SAPs. The SAPs calculate their budgets for all channels. (The auction result will be announced in stage III.) In stage II, the SH sells channels to SAPs. In stage III, the winning SAPs determine winning SUs, serve and charge them. TASG meets the needs of spectrum group-buying for SUs. It is shown in Fig. 4.1.

4.3.1 *Stage I: Group Budget Calculation for the Outer Auction*

We first introduce the algorithm to calculate the budget for SAPs, which is the crucial factor to form SAPs' best bids in stage II. Then we give a simple example for illustration. In this stage, preparation works will be done for real trade in the last two stages.

4.3.1.1 Algorithm for Calculating the Budget Vector for SAPs

For simplicity, we assume winning SUs will equally share the channel and payment. So the winning SUs will get the same capacity and pay the same amount.

First, SAPs obtain characteristics of channels from the SH, such as the bandwidth, frequency, and power limitation. These properties will be passed to SUs to help them make accurate evaluations and form their best strategies.

SAP P_j will receive the two-dimension bids for channels from its SUs: $\{(v_i^j(k), d_i^j(k))\}$, for all i and k. In this stage we design an algorithm to calculate the budget vector of channels $\{H_k\}$ for SAP P_j. We divide the SUs into two subsets: winner set $\mathbb{W}_j$ and loser set $\mathbb{L}_j$. The clearing price of a channel should be a budget value selected from the loser set to prevent untruthfulness. This follows the design of previous works [11,13]. To become a winner, an SU should satisfy two conditions.

1. Its budget should be no less than the clear price c, which is the payment for all winners in this SN ($\forall S_i^j \in \mathbb{W}_j, \dot{d}_i^j(k) = c$). Otherwise it cannot afford the clearing price and should be assigned to loser set.
2. Its evaluation of the channel should be high enough. Specifically, it should be at least the product of the number of winners and the clearing price. We will explain this condition later.

In order for the mechanism to achieve truthfulness, Part of the social welfare should be sacrificed, e.g. reduce the number of transactions. VCG auctions charge buyers low prices thus reduce system efficiency. Double auctions hurt a couple's (a buyer's and a seller's) interests as their bid and ask are selected as market clearing prices [11, 13]. In this work these sacrificed users will not be the winning ones if SAP P_j is charged H_k (in line 9) by SH.

SAMU is short for *Sacrificing At least M Users*, which is proposed to obtain the budget vector. The general idea of SAMU is to remove those SUs with lower budget and evaluations. From SAP, an integer m is randomly generated. Then m SUs with the smallest budgets are sacrificed. Finally, the SAP goes through the remaining SUs and removes those with small enough evaluations.

Algorithm 4 SAMU: SAP P_j Calculating the Budget Vector for Channels

for $k = 1$ to C **do**
  Let $\mathbb{T}$ be the sorted array of SUs in descending order of budget $\{d_i^j(k)\}$.
  Let $1 \leq m \leq N_j - 1$ be a bid-independent integer.
  Let $\mathbb{T}'$ be the sorted array of $\mathbb{T}$'s first $N_j - m$ elements, in descending order of evaluations.
  Let clearing price c be the boundary SU $\mathbb{T}_{N_j-m+1}$'s budget.
  for $l = N_j - m$ to 1 **do**
    if $\mathbb{T}_l'$'s evaluation is no less than $l \times c$ **then**
      $H_k = l \times c$, with $\mathbb{T}_l'$'s first l SUs forming the winner set $\mathbb{W}_j$
      break
    end if
  end for
end for
return vector $\{H_k\}$

Restriction of $m = 1$ reduces the clearing prices and thus the probability that the SAP wins a channel. Instead of sacrificing only one SU or one group. SAMU makes the selection of m more flexible to improve the group budget. If not restricted to $m = 1$, the budget H_k can usually increase. When there are more than one users/groups submitting relatively much lower budget than others, the potential market clearing price will be dragged down by $m = 1$. A lower clearing price can reduce the possibility of SAP's winning a channel if it is smaller than sellers' reserved prices. Once the SAP wins a channel, it guarantees non-negative utility for its SUs as well as itself, while failing to obtain a channel gives zero utilities for all its SUs and itself. If there are many SUs and their bids randomly distributed in a region, SAMU with $m > 1$ performs better than that with $m = 1$. We will show the advantage of SAMU in the Sect. 4.5.

Note that SAMU calculates the SAP's budget and corresponding winner set for each channel. But till now the SAP does not know whether it can win a channel from the SH and which channel it will obtain. So it cannot announce the auction result immediately.

Actually can treat H_k as SAP P_j's true evaluation as well as budget of the k-th channel. H_k is the maximum amount it can charge those winning SUs. If the SH charges SAP P_j more than H_k for trading the k-th channel, it will definitely received a negative utility eventually, unsatisfying individual rationality and budget balance properties.

4.3.1.2 A Simple Example

Here we give some simple examples to further illustrate the idea of SAMU.

In the example shown as Fig. 4.2, there are 8 SUs in the SN. SAP P_j chooses $m = 2$. Each solid line connects the true evaluation and budget values for the same channel from the same SU. First, P_j sorts the budget, selects 24 as clearing price

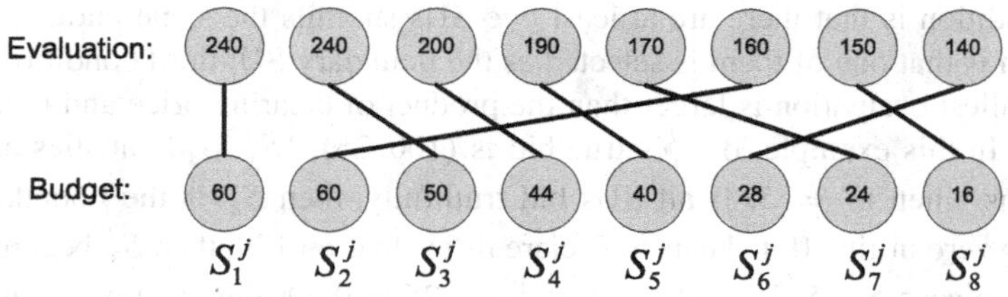

Fig. 4.2 Illustration of SAMU

and removes S_7^j and S_8^j. Then P_j sorts the remaining 6 SUs by their evaluations. SU S_6^j is removed as $140 < 6 \times 24$. Then SAP finds that $160 \geq 5 \times 24$. So $H_k = 120$.

S_6^j is willing to pay (though it does not have) 140 units of money to buy the whole channel. If it participates the group-buying, it will share $\frac{1}{6}$ of the whole channel, which is worth $\frac{1}{6} \times 140 < 24$ units of money for itself. So it is not willing to participate the group-buying at price 24. This explains the second condition of a winning SU.

If $m = 1$, then $H_k = 7 \times 16 = 112 < 120$. In some extreme cases with some SUs submitting much lower evaluations or budgets, the competitiveness of this SN will be significantly reduced. Those SUs with lower evaluations or budgets cannot share a channel anyway. Worse still, they become a burden to other SUs submitting higher values. So we believe that being less restrictive on the number of sacrificed users in an SN can help improve the system efficiency and fairness, which is good for everyone.

Although a high budget H_k is desirable, we discover that if SAP P_j tries to maximize H_k (equivalently, find the optimal m value), it can lead to SUs' untruthful bidding. So m should be randomly selected, and independent of the current bids.

To illustrate SUs' untruthful bidding, we enumerate all the possible values of $m = \{1, \cdots, 7\}$, and find the corresponding $H_k = \{112, 120, 140, 160, 132, 100, 60\}$. Selecting the winner set $\mathbb{W}_j = \{S_1^j, \cdots, S_4^j\}$ with clearing price 40 maximizes H_k. If all SUs submit truth values, S_4^j is winner with utility $\frac{200}{4} - 40 = 10$. If others keep their submissions and S_4^j submits untruthful value (200, 30), then the new highest H_k is 140 with $\mathbb{W}_j = \{S_1^j, \cdots, S_5^j\}$ and clearing price 28. S_4^j's utility will be $\frac{200}{5} - 28 = 12 > 10$. S_4^j's true bid is not a dominant strategy. See through the appearance to the essence, if the size of winner set depends on SUs' bids in order to maximize H_k, some winner may be able to cheat to change the winner set, such that he is still in the new winner set and gets his utility improved. If independently selected, any winner who wants to change the winner set should submit a lower budget and become a loser. Obviously, he has no incentive to do this.

The major defect of SAMU is that in some cases that seldom happen, there may be untruthful bids. Three conditions must be satisfied to make SAMU invalid.

First condition is that there are at least two SUs submits the same budget; second condition is that one of them is selected as the boundary SU; third condition is that their smallest evaluation is larger than the product of clearing price and number of winners. In this example, if S_4^j's true bid is $(100, 25)$, $\{S_3^j, S_4^j\}$ satisfies all three conditions when $m = 5$. If all SUs bid truthfully, then S_4^j is the boundary SU, obtaining zero utility. If it cheats to declare its evaluation 130, then S_3^j become loser and S_4^j replaces S_3^j. S_4^j's utility will be $\frac{100}{3} - 25 > 0$. Originally, this is caused by the two-dimension structure of SUs' bids. Given the SUs do not collude and their true values randomly distributed in a continuous region, the probability that at least two SUs submit their budget of the same value is zero. It means that it is almost impossible to satisfy the first condition and SAMU is valid for almost all cases.

4.3.2 *Stage II: The Inner Auction for Distributing Channels to SAPs*

By now the SAPs have calculated their budget vector $\{H_k\}$. In this part, we will describe how the trading is made between SAPs and the SH. We first give Algorithm 5, DCP (Diverse Clearing Prices) to enable the inner auction and then an illustrative example.

4.3.2.1 DCP: Algorithm for Distributing Channels to SAPs

What we are about to do here is to design an auction to trade the channels with respect to the desired economic properties we hope to achieve, given $\{b_j^k\}$ and $\{r_k\}$. Note that each SAP P_j submits a bid vector $\{b_j^k\}$ to the SH. The SH can construct a bid matrix $\{b_j^k\}$ (for all j, k) consisting of the bid vectors. The SH also has a vector of reserved prices for all channels $\{r_k\}$.

This auction is a single seller, multiple buyers, multiple heterogeneous items auction. McAfee mechanism based auctions [8, 11, 13] do not work as they are designed for homogeneous items. It is well known that VCG auctions are extremely computationally hard, therefore we also cannot directly apply VCG auction here. The time required in VCG auction is exponential of the input size of the problem. VCG double auction can be excluded as it does not satisfy the budget balanced property. We cannot force auctioneers to receive a negative utility. Another candidate is the TASC-like mechanism [5, 12], which is originally designed for double auctions which requires sellers and buyers to act truthfully. Whilst the problem in this stage is a single auction, which does not restrict so much on the seller. If applying TASC without many modifications, the system efficiency is low and unsatisfying.

Algorithm 5 DCP: SH's auction for distributing channels to SAPs

SH randomly matches SAPs with channels. Make sure that the matching result **m** is independent of the bidding vectors of SAPs. Denote the set of matched pair of a SAP and a channel as $\mathbb{S}$.
Let vector $\mathbb{B}'$ be the sorted $\{b_j^k\}$ in descending order.
for all $(P_j, k) \in \mathbb{S}$ **do**
  if $b_j^k \geq r_k$ **then**
    use binary search to find b_j^k and r_k in $\mathbb{B}'$. We get the index x such that $\mathbb{B}'_{x-1} > b_j^k \geq \mathbb{B}'_x$, and the index y such that $\mathbb{B}'_y \geq r_k > \mathbb{B}'_{y+1}$. (If b_j^k is the largest element in $\mathbb{B}'$, $x = 1$. If r_k is less than any elements in $\mathbb{B}'$, then $y = |\mathbb{B}'|$.)
    for $l = y$ to x **do**
      if element $\mathbb{B}'_l$ is submitted by P_j **then**
        continue
      else
        SAP P_j is allocated the k-th channel. Its clearing price is $\dot{b}_j^k = \mathbb{B}'_l$.
        break
      end if
    end for
  end if
  if SAP P_j fails to obtain k-th channel **then**
    SAP P_j fails to obtain any channel.
  end if
end for

To solve the shortcomings in existing works, we design DCP algorithm for the SH to distribute channels to the SAPs. In the design of DCP, the SH first generates a random Maximum Matching (MM) **m**, and then check each pair (of the k-th channel and SAP P_j) in **m**. If SAP P_j's bid b_j^k is less than the reserved price r_k, it fails to obtain the k-th channel, in fact, any other channels. Otherwise, we search through all the bids. If there is some bid(s) submitted by other SAPs within the region $[r_k, b_j^k]$, then SAP P_j can obtain the k-th channel. Its clearing price is the lowest bid within $[r_k, b_j^k]$ from other SAPs. If there is no such bid, SAP P_j fails to obtain a channel.

A random MM is the best choice the SH can make. MM is independent of SAPs' bids, so it works. We are also aware of the fact that the matching result **m** has a close relationship with the system efficiency. A Maximum Weight Matching (MWM) is able to optimize the system efficiency. But we can show that MWM is likely to lead to untruthful bids from SAPs. Because MWM makes allocations which are dependent of SAPs' bids. The reason is similar with the one that a SAP cannot try to maximize its budget H_k in stage I. It is possible that some SAP submits a fake bid to manipulate the matching result, thus obtains improper benefits. So it is necessary to make the matching result independent of the bids.

TASC-like mechanism restricts the auctioneer to decide a Uniform Clearing Price (UCP) for all winners and channels. (So we imitate TASC-like mechanism with UCP algorithm in this work, which uses the same price to clear the market.) DCP performs no worse than UCP in this stage, given the same matching result

Table 4.2 Bid matrix

	1st channel	2nd channel	3rd channel
SAP P_1	100	18	9
SAP P_2	81	30	9
SAP P_3	85	25	12 (0)

is selected. We believe uniform clearing pricing is one of the critical techniques to make the heterogeneous items double auction truthful. But the problem in this stage is not double auction. Besides, it does not come very naturally that the heterogeneous channels should be sold in the same price, especially when the channels are quite different in the qualities such as bandwidth, frequency and noise level. DCP allows heterogeneous channels to be traded at different prices, which will promote the trading volume and improve the system efficiency. Because if a trade happens under UCP, we are also able to find a clearing price for this pair (at least the price that the SAP pays under UCP will be selected by DCP), given the same matching result.

4.3.2.2 A Simple Example

We give a simple example in this part. It shows that if the matching result is dependent on SAPs' bids, it is possible for some SAP to manipulate the auction to obtain improper benefits. Then we show the process and result of DCP, and compare its performance with UCP.

Consider the bid matrix shown in Table 4.2, where all SAPs submit their true evaluation. The reserved price vector is $\{80, 16, 8\}$.

Now we show the process and result of DCP. We assume the matching result is $\mathbf{m} = \{(P_1, 3), (P_2, 2), (P_3, 1)\}$. The vector $\mathbb{L} = \{100, 85, 81, 30, 25, 18, 12, 9, 9\}$. For $(P_1, 3)$, $x = 8, y = 9$. We select $b_2^3 = 9$ as the clearing price for P_1. For $(P_2, 2)$, $x = 4, y = 6$. We select $b_1^2 = 18$ as the clearing price for P_2. For $(P_3, 1)$, $x = 2, y = 3$. We select $b_2^1 = 81$ as the clearing price for P_3.

Then we show that a bid-dependent matching result can lead to untruthfulness. If using MWM, the SH will choose the pair $\{(P_1, 1), (P_2, 2), (P_3, 3)\}$. Let us focus on SAP P_3. It will obtain 3rd channel at the price of 9. Its utility is $12 - 9 = 3$ (assume it will charge its SUs an aggregate amount of 12 later). If SAP P_3 submits a fake value of 0 instead of 12 for the 3rd channel, the MWM result will be $\{(P_1, 1), (P_2, 3), (P_3, 2)\}$. Now it obtains 2nd channel and pays 18. Its utility will be $25 - 18 = 7$, which is higher than 3 when it bids truthfully.

If applying UCP with the same $\mathbf{m}$, no transaction can be executed. In fact, DCP preforms better if the channels are of high diversity while UCP wastes too many the trading opportunities.

4.3.3 *Stage III: The Outer Auction's Winner Determination and Pricing*

In this part, we discuss how a winning SAP determines the winning SUs, serves and charges them. We assume SAP P_j obtains the k-th channel at price $\dot{b}_j^k$. By our algorithms, $\dot{b}_j^k \leq H_k$.

SAP P_j should announce those SUs who contribute an aggregate amount of H_k as the winners and charge them exactly H_k. It means that the real allocation and payment within this SN is exactly the same as the preparation result in the first stage. P_j will receive a non-negative utility $U_j = H_k - \dot{b}_j^k \geq 0$.

Note that SAP P_j should not select any other set of SUs as winner set even if they also offer budget higher than $\dot{b}_j^k$. P_j also should not charge winner set less than H_k even if $\dot{b}_j^k$ is strictly less than H_k. Otherwise it is very likely that some SUs have incentive to cheat. Counter-examples can be easily given. In a word, the winner set determination and pricing should strictly follow the rules to guarantee the truthfulness of TASG.

For example, in Fig. 4.2, with $m = 4$, the winner set $\{U_1^j, U_2^j, U_3^j, U_4^j\}$ gives budget 160. We assume the SH charges SAP P_j 120 eventually. P_j should insist on the same winner set. Each winner will be charged $\frac{160}{4} = 40$ and share one quarter of the channel. P_j's utility U_j is maximized to be $160 - 120 = 40$. The four winners' utilities are $\frac{240}{4} - 40 = 20$, $\frac{160}{4} - 40 = 0$, $\frac{240}{4} - 40 = 20$ and $\frac{200}{4} - 40 = 10$ respectively.

4.4 Economic Properties and Time Complexity

In this section, we prove that TASG satisfies the truthfulness and analyze its time complexity. Individual Rationality and Budget Balance is self-evident through the design process and analysis of our algorithms. So we do not discuss these two properties here.

4.4.1 *Truthfulness*

Theorem 4.1. *TASG is truthful (at both inner auction and outer auction).*

The proof is established on a series of lemmas. Due to page limitation and the similarity with our previous work [5], we do not provide details for some of the proofs.

The statements and proof are based on the precondition that only one player adjust its strategies while all other players' strategies and rules are fixed. This follows the definition of truthfulness and is consistent with previous works.

Lemma 4.1. *If SU $S_i^j(k)$ wins both when submitting $(\tilde{v}_i^j(k), \tilde{d}_i^j(k))$ and $(v_i^j(k), d_i^j(k))$, it pay the same price. If SAP P_j wins both (the k-th channel) when submitting H_k and b_j^k, it pays the same price.*

Lemma 4.2. *If SU $S_i^j(k)$ wins when submitting its true values $\tilde{v}_i^j(k)$ and $\tilde{d}_i^j(k)$, it can also win by submitting any other $v_i^j(k) \geq \tilde{v}_i^j(k)$ and $d_i^j(k) \geq \tilde{d}_i^j(k)$. If SAP P_j wins (the k-th channel) when submitting its true value H_k, it can also win by submitting $b_j^k > H_k$.*

Proof. For these two lemmas in both the stages, the proof is similar as that in [5]. Because a higher bidding value (or evaluation, budget) will not make the position of the SAP (or SU) moved along the descending direction in the algorithms. Given the rules and others' strategies unchanged, the winner set and the boundary price will keep the same. So the SAP (or SU) is still winner and pays the same amount.

Lemma 4.3. *TASG is two-dimensional truthful in $(v_i^j(k), d_i^j(k))$ for SU S_i^j.*

Proof. We discuss all the four cases to show that the SU cannot improve its utility by submitting untruthful values.

1. Case 1: S_i^j fails when truthful and wins when untruthful. By SAMU, when truthful, S_i^j fails either because $\tilde{d}_i^j(k)$ is smaller than the clear price c, or $\tilde{v}_i^j(k)$ is smaller than the product of c and number of winners $|\mathbb{W}_j|$. If $\tilde{d}_i^j(k) < c$, then it has to submit $d_i^j(k) \geq c$ to win. The clear price c will not decrease in this case. So S_i^j wins and needs to pay at least c. It makes his utility zero as the amount exceeds his real value $\tilde{d}_i^j(k)$. Otherwise, we have $\tilde{d}_i^j(k) \geq c$ and $\tilde{v}_i^j(k) < c|\mathbb{W}_j|$. S_i^j has to submit $v_i^j(k) \geq c(1 + |\mathbb{W}_j|)$ to win. So the profit it gets $\frac{\tilde{v}_i^j(k)}{1+|\mathbb{W}_j|}$ is less than its payment c.
2. Case 2: S_i^j wins when both biding truthfully and untruthfully. By the above lemmas, it pays the same. Its untruthful strategies should satisfy $d_i^j(k) \geq c$ and $v_i^j(k) \geq c|\mathbb{W}_j|$. So its strategy does not affect the winner set. It has to share the channel with the same other winners.
3. Case 3: S_i^j wins when truthful and fails when untruthful. By our rule, winner's utility is non-negative and loser's utility is zero.
4. Case 4: S_i^j fails when both biding truthfully and untruthfully. So its utility is always zero.

The above analysis proves the truthfulness of SU S_i^j if there is only one channel for them to select. We take this only channel as the one being matched to SAP P_j. For other channels, no matter S_i^j cheats or not on their evaluations or budgets, it makes no differences.

If there are some mapping relationships between a SU's bid values and the state II matching result **m**, then the SU may be able to cheat to influence **m** to improve its utility. So we see that the independence of **m** is much critical to guarantee the truthfulness of the two stages.

Lemma 4.4. *TASG is truthful for SAPs. (Any SAP P_j's best bid vector is its $\{H_k\}$.)*

Proof. For SAPs' auction, the matching result **m** decides the assignment of SAP P_j and the k-th channel. If P_j fails to obtain this channel, it has no more opportunities. The proof structure is to similarly discuss the four cases.

4.4.2 *Time Complexity*

We now analyze the running time of two algorithms in TASG.

Theorem 4.2. *The overall time complexity is $O(CN_j \log N_j)$ for SAMU algorithm, and $O(min\{N^2C, NC^2\})$ for DCP algorithm.*

Proof. For SAMU, sorting needs $O(N_j \log N_j)$ time. The inner for-loop takes $O(N_j)$ time.

For DCP, generating a random match needs $O(min\{N, C\})$ time. Sorting takes $O(NC \log(NC))$ time. Binary search needs $O(\log(NC))$ time. The inner for-loop takes at most $O(NC)$ time.

4.5 Evaluation

In this part we present simulation results to verify our conclusions , evaluate the performance and compare it with existing mechanisms. The experiment environment is MATLAB. We will show that TASG is truthful for SUs and SAPs first. Then we evaluate SAMU under different m values, and compare DCP with UCP in two stages, respectively. Except for Figs. 4.3 and 4.6, other data have been averaged over 100 cases with randomly generated parameters.

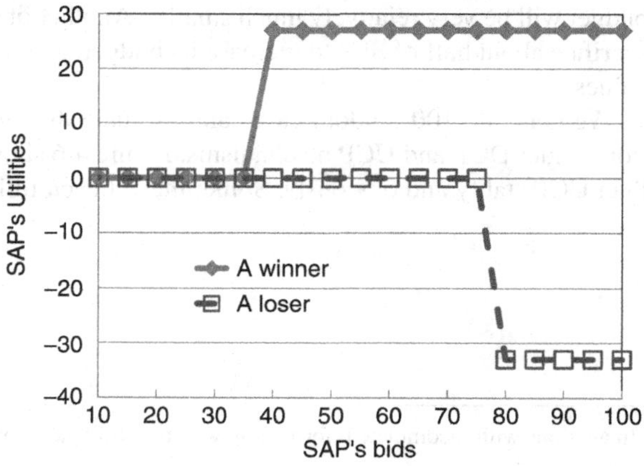

Fig. 4.3 Truthfulness of DCP

Table 4.3 Truthfulness of SAMU

m	$U_i^j(u) < U_i^j(t)$	$U_i^j(u) = U_i^j(t)$	$U_i^j(u) > U_i^j(t)$
1	0.4133	0.5867	0
2	0.3907	0.6903	0
4	0.2349	0.7651	0

The default setting of parameters are $C = 10$, $N = 10$, $N_j = 100$ for $j = 1, \cdots, N$. $\tilde{d}_i^j(k)$ and $\tilde{v}_i^j(k)$ are uniformly distributed within $[0, 10]$ and $[0, 30]$ respectively for stage I. b_j^k and r_k are uniformly distributed within $[50, 100]$ and $[40, 80]$ respectively for stage II.

We randomly select an SU S_i^j, adjust its budget and evaluation of the k-th channel within the range of $[0, 2\tilde{d}_i^j(k)]$ and $[0, 2\tilde{v}_i^j(k)]$ respectively. Table 4.3 shows the truthfulness of SAMU. $U_i^j(u)$ and $U_i^j(t)$ are SU S_i^j's utilities when he bids untruthfully and truthfully respectively. The values are the probabilities of the three cases with different m value. We see that in any case, $U_i^j(u) > U_i^j(t)$ cannot happen, which supports the truthfulness of SAMU. With larger m values, the probability that $U_i^j(u) < U_i^j(t)$ is rising, as the number of losers increases. No matter the losers cheat or not, they are more likely to get the same utilities zero.

Figure 4.3 verifies the truthfulness of DCP. We select two typical SAPs, one winner and one loser. The winner achieves positive utility if it bids truthful. If it submits low bid, it may fail and get zero utility. The loser achieves zero utility if it bids truthfully. If it submits high bid, it can win a channel but its utility will be negative.

Figure 4.4 plots the utility of the SH with different number of SAPs. The SH's utility decreases with the number of SAPs. We can call it "seller's curse". This is an interest phenomenon. With the number of SAPs increasing, the "winner's curse" [1] is reduced and the "seller's curse" becomes severe as the winners pay the smallest amount of money.

Figure 4.5 compares the performance of different m values of SAMU . We see that when budget and evaluation are uniformly distributed, there may be quite a few SUs with low budget and evaluations. If sacrificing only one of them, the SAP's budget will be very relatively much smaller. Another observation is that the SAP can sacrifice about half of SUs to increase its budget, given their uniform distribution of values.

We generate 100 random cases and evaluate the number of successful transactions under DCP and UCP mechanisms. Figure 4.6 shows that DCP performs better than UCP stably and obviously. Sometimes, it even fully execute all the deals.

[1] In auctions with incomplete information, winners will tend to overpay.

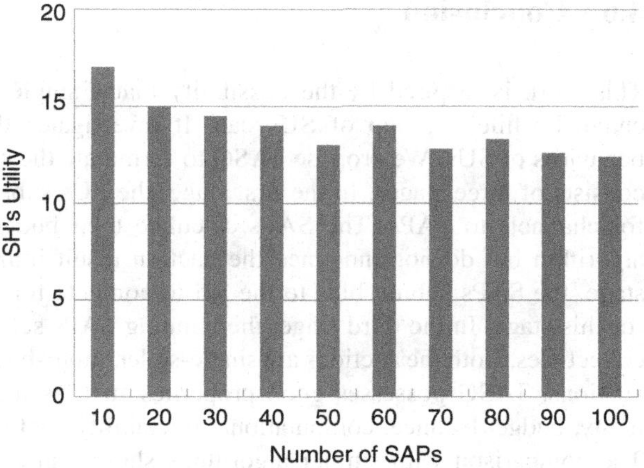

Fig. 4.4 SH's utility

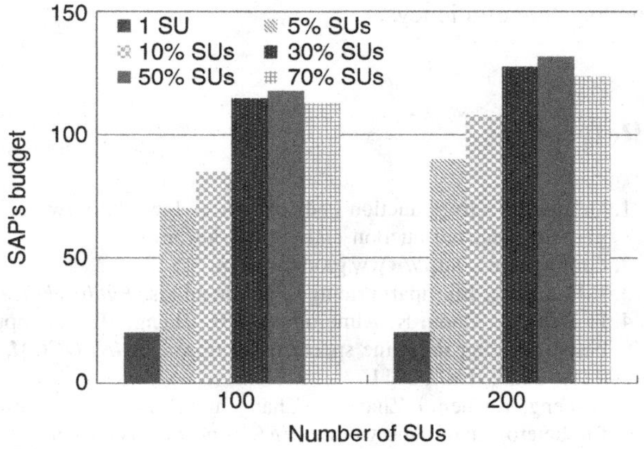

Fig. 4.5 Comparison of different *m* values of SAMU

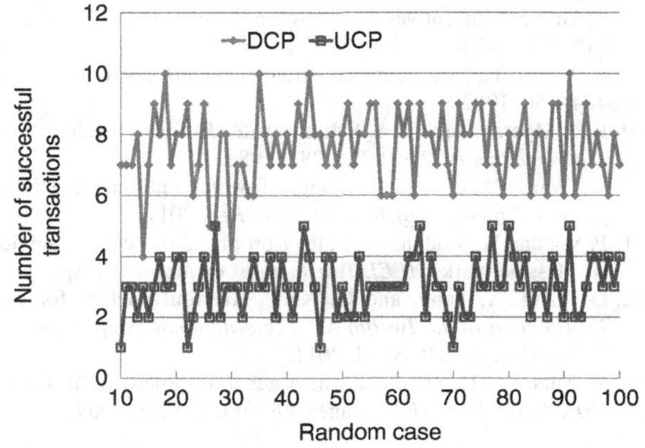

Fig. 4.6 Comparison of DCP and UCP

4.6 Conclusion

This work is inspired by the possibility that a single SU cannot afford a whole channel while a group of SUs can. It investigates the spectrum group-buying behaviors of SUs. We propose TASG to formulate the scenario and solve it. TASG consists of three stages. In the first stage, the SUs submit budgets and evaluations for channels to SAPs. The SAPs calculate their budgets of channels by SAMU algorithm but do not announce the auction result immediately. In the secondary stage, the SAPs submit bids to the SH to compete for channels. DCP is proposed for this stage. In the third stage, the winning SAPs select winners, serve them and collect fees. Both the auctions are single-seller, multi-buyer, heterogeneous channels auctions. TASG possesses good properties such as truthfulness, individual rationality, budget balance, computational tractability, and improved system efficiency. The comparison with similar algorithms shows our advantages theoretically. We also illustrate with examples to give intuitive impressions to help understand the algorithms. Simulation results verify our theoretical conclusions on the truthfulness and system efficiency.

References

1. German spectrum auction ends but prices low. http://www.rethinkwireless.com/2010/05/21/german-spectrum-auction-ends-prices-low.htm.
2. Groupon. inc. http://www.groupon.com.
3. E.H. Clarke. Multipart pricing of public goods. *Public choice*, 11(1):17–33, 1971.
4. L. Deek, X. Zhou, K. Almeroth, and H. Zheng. To preempt or not: Tackling bid and time-based cheating in online spectrum auctions. In *INFOCOM, 2011 Proceedings IEEE*, pages 2219–2227. IEEE, 2011.
5. X. Feng, Y. Chen, J. Zhang, Q. Zhang, and B. Li. Tahes: A truthful double auction mechanism for. heterogeneous spectrums. *IEEE Transactions on Wireless Communications*, 2012.
6. X. Feng, J. Zhang, and Q. Zhang. Database-assisted multi-ap network on tv white spaces: Architecture, spectrum allocation and ap discovery. In *DySPAN, 2011 IEEE Symposium on*, pages 265–276. IEEE, 2011.
7. T. Groves. Incentives in teams. *Econometrica: Journal of the Econometric Society*, pages 617–631, 1973.
8. R.P. McAfee. A dominant strategy double auction. *Journal of economic Theory*, 56(2): 434–450, 1992.
9. R.B. Myerson and M.A. Satterthwaite. Efficient mechanisms for bilateral trading. *Journal of economic theory*, 29(2):265–281, 1983.
10. W. Wang, B. Li, and B. Liang. Towards optimal capacity segmentation with hybrid cloud pricing. *University of Toronto, Tech. Rep*, 2011.
11. F. Wu and N. Vaidya. A strategy-proof radio spectrum auction mechanism in noncooperative wireless networks. *IEEE Transactions on Mobile Computing*, 2012.
12. D. Yang, X. Fang, and G. Xue. Truthful auction for cooperative communications. In *Proceedings of the Twelfth ACM International Symposium on Mobile Ad Hoc Networking and Computing*, page 9. ACM, 2011.
13. X. Zhou and H. Zheng. Trust: A general framework for truthful double spectrum auctions. In *INFOCOM 2009, IEEE*, pages 999–1007. IEEE, 2009.

Chapter 5
Flexauc Auction: Serving Dynamic Demand in Wireless Market

In wireless market, major operators buy spectrum through auctions held by spectrum regulators and serve end users. How much spectrum should an operator buy and how should he set the optimal service tariff to maximize his own benefits are challenging and important research problems. On one hand, a Wireless Service Provider's (WSP's) strategies in spectrum auction and service provision are coupled together. On the other hand, spectrum holder wants to design an auction to flexibly satisfy operators' spectrum requirement, improving sales revenue and maximize social welfare. Previous works do not see the big picture and usually study one of the sub-problems.

In this work, we jointly study the strategy of the SH in the auction design and the WSPs' strategies in the service provisions and biddings. The WSP's optimal strategy in the auction can be flexible in term of demands and valuations. To optimize social welfare and enable the WSPs to reveal truthful flexible demands, we design Flexauc, a novel auction mechanism for the SH. We prove theoretically that Flexauc not only maximizes the social welfare but also preserves other nice properties: truthfulness and computational tractability.

5.1 Bid According to Dynamic Demands

In this chapter, we will concentrate on flexible spectrum auction in primary market to serve the dynamic demand from users. The key factors in this design are the same as spectrum group-buying framework, such as truthfulness, efficiency and computational complexity.

Auctions are usually applied for spectrum redistribution to increase efficiency. In the spectrum trading markets, major wireless service providers (WSPs) purchase spectrum through auctions organized by the spectrum holders (SH). It is essential for the WSPs to determine proper bidding strategies in the auction to optimize the revenues. We observe that the WSPs' bidding strategies relate tightly to the service provisions to the end users. Both the WSPs' demands and valuations can be flexible in the auction due to their end users' heterogeneous demands and willingness to pay. Considering these flexibilities, it is therefore important for the SH to design auction schemes to not only optimize social welfare but also enable the buyers to reveal their truthful demands and valuations.

We observe two facts. First, the operators' spectrum demands directly depend on their users' demands. Second, their willingness to pay for the spectrum is also dependent on their pricing schemes for the users. That is to say when participating in a spectrum auction, to form his best bidding strategy in terms of both the quantity to buy and the claimed price, an operator should consider his users' responses.

However, WSPs' flexible bidding strategies cannot be satisfied by existing works. First of all, there are no suitable existing auction schemes that are able to promote flexible biddings with computational efficiency. Most works rely on over-simplified assumptions or induce heavy overheads. Some of them [6, 13, 14] assume that one buyer can claim for at most one channel. Some others [8, 11] assume that a buyer can submit bids for multiple channels but win at most one. These assumptions indeed limit the scope of applications especially when a buyer is willing to win more than one channels. Although combinatorial auctions (e.g. [3]) can meet the requirements of flexible auction, it induces very heavy computational overheads. Moveover, as far as we know, no work has studied the relationship between users' willingness to pay and WSP's true evaluation of the channels. In the auction design, buyers and sellers have values in their minds for the commodity, which are defined as the true evaluations. It is a quantitative measurement of the satisfaction of consumption behavior. This concept is the core in any auction design. But no existing works study how an operator determines its evaluation of the channels.

This chapter aims to address the two above-mentioned problems. Considering the service provision between the WSPs and their end users, a truthful auction mechanism for the SH to enable bidders' flexible strategies in terms of demands and valuations and also preserves computation efficiency is designed. Specifically, a three-layered spectrum trading scenario shown in Fig. 5.1 consisting of the SH, the WSPs and the end users is considered. In the spectrum auction between the SH and the WSPs, the SH partitions the spectrum into channels with equal bandwidth for sale and designs the auction mechanism. The WSPs then bid the channels from the SH. In the service provision between the WSPs and the end users, the WSPs decide their optimal service prices for their end users and the users determine how much bandwidth they want to consume. Different from most existing works considering such a multi-layered spectrum trading markets [4, 12], in this work, auction instead of pricing is conducted between the SH and the WSPs. The pricing mechanism is often used when the sellers know precisely the value of selling resources. The auction mechanism on the other hand assumes no such information, which suits our scenario better.

In this work, we design a novel auction mechanism referred to as Flexauc for the WSPs' flexible bidding strategies. With Flexauc, the bidders can bid for any amount of demands with different valuations. It is an improved version of the combinatorial auction for identical items with significantly reduced computational overheads. In Flexauc, each WSP submits a series of bids representing his marginal benefit of

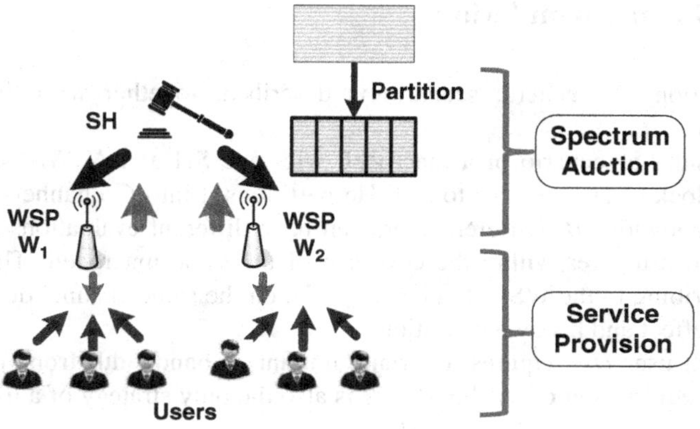

Fig. 5.1 A three-layered network scenario

the additional channels. A WSP is possible to win any number of channels and he will be satisfied with any possible result. We model the strategic behaviors in each layer and leverage backward induction to compute the WSPs' bids in the auction. We analyze the WSPs' true valuations for the channels based on their pricing strategies and users' possible responses in the service provision. Note that existing works usually assumes the valuations are fixed and known inputs for the auction problem. In Flexauc, we propose two new payment mechanisms which are proved to be truthful and maximize social welfare.

In summary, the key conclusions and contributions of this work are summarized as follows.

1. By considering both the auction and service provision in wireless market, we propose a framework for the entire market. We design Flexauc, and show its advantages over previous mechanisms.
2. This work studies the flexible auction mechanism, where the operators can flexibly decide the best quantity of channels to buy and the best bidding values, as well as the service price for end users.
3. We present comprehensive simulation results to verify our theoretic conclusions.

The rest of this work is organized as follows. The detailed system model are described in Sect. 5.2. In Sect. 5.3, we discuss the spectrum partition, and elaborate auction mechanism, bidding behaviors and the demand response as well as pricing mechanism. We analyze the economic properties in Sect. 5.4, specifically truthfulness, efficiency and time complexity. Section 5.5 provides our numerical results to verify our design objectives and make comparison with previous works to show advantages of Flexauc. We conclude this work in Sect. 5.6.

5.2 Problem Formulation

In this section, the general scenario is described, together with the problem formulations.

We consider a scenario of a macrocell with one SH and N WSPs. SH has a spectrum block of size $B \times C$ to sell. He partitions it into C channels, each with the same bandwidth B. Different users can have different evaluations Each WSP deploys infrastructures within the coverage of the same macrocell. There are N_i users subscribing to the WSP W_i $(1 \leq i \leq N)$. on the same channel due to fading, multi-path effect and his own conditions.

W_i's j'th user U_{ij} requires a certain amount of bandwidth from W_i. Let the bandwidth requirement of U_{ij} be w^i_j. w^i_j is also the only strategy of a user. U_{ij} can achieve a data rate of $\alpha_i w^i_j \ln(1 + \frac{g^i_j}{w^i_j})$, where α_i is his evaluation of the data rate and $g^i_j = \frac{P_i H_j}{n_0}$. H_j is the path loss factor. n_0 is the power density of thermal noise. P_i is the transmission power level of base stations of W_i. Let p_i be the unit price of the shared bandwidth charged by W_i. So U_{ij}'s payment is $p_i w^i_j$. U_{ij}'s utility can be defined as:

$$U^u_{ij} = \alpha_i w^i_j \ln(1 + \frac{g^i_j}{w^i_j}) - p_i w^i_j. \tag{5.1}$$

A WSP W_i's bid is defined as $\{b_{i1}, \cdots, b_{iC}\}$. The meaning of the bid is that W_i is willing to pay b_{i1} for its first allocated channel, b_{i2} for his second channel, and so on. If W_i wins K_i channels, we define the utility as the difference between the revenue from users and the cost of channels:

$$U_i = p_i \cdot \min\{\sum_{j=1}^{N_i} w^i_j, K_i B\} - \sum_{j=1}^{K_i} \tilde{b}_{ij}, \tag{5.2}$$

where $\tilde{b}_{ij}$ is W_i's payment to SH for the j-th channel. $\tilde{b}_{ij} \leq b_{ij}$ should hold as the payments for the channel should be less than the WSP's bid value. In the equation, p_i and $\{b_{ij}\}$, $(1 \leq j \leq C)$ are W_i's strategies. The auction results, K_i and $\{\tilde{b}_{ij}\}$, are determined by SH.

We define the utility of SH as his revenue from spectrum sales, which can be written as:

$$U_{SH} = \sum_{i=1}^{N} \sum_{j=1}^{K_i} \tilde{b}_{ij}. \tag{5.3}$$

A typical process of the scenario is as follows: first SH makes spectrum partition and rules of auction, and W_i evaluates the demand requirements of users given its price p_i. By the evaluation, W_i can calculate its optimal bidding value

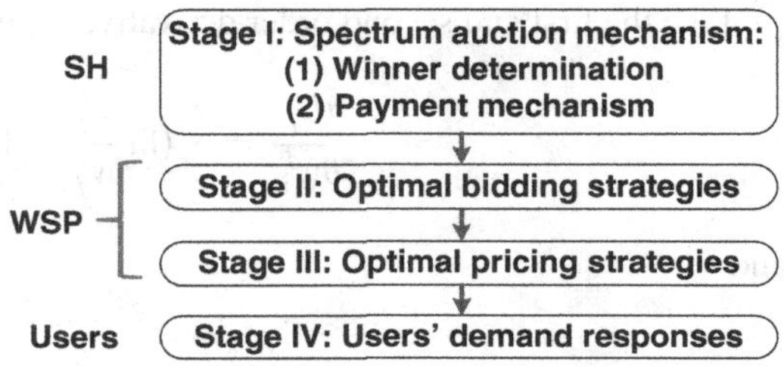

Fig. 5.2 A four-staged analysis framework

$\{b_{i1}^*, \cdots, b_{iC}^*\}$. Then SH collects all the bid values, and determines winners and the winners' payment ($\{K_i\}$ and $\{\tilde{b}_{ij}\}$). After W_i gets informed K_i and $\tilde{b}_{ij}$, it announces to the users his optimal price p_i^*.

Note that the related concepts we use in this chapter are the same as those in Chap. 2. Specifically, we consider the aggregate utility of the SH and WSPs as social welfare, denoted by $S = \sum_{i=1}^{N} U_i + U_{SH}$.

5.3 Flexible Auction Design

In this section, we make theoretical analysis for this problem in the scenario shown in Fig. 5.2. There are four stages in the entire procedure. First, the SH decide channels to sell and the auction mechanism. Second, we deduce the optimal bidding strategies of WSPs. Third, we derive the optimal pricing strategy for service tariff. In the last stage, we analyze the best demand strategy for users. Backward induction is leveraged for the analysis. The results from the latter stages will be first analyzed and serve as input for earlier stages. At the end of the section, we will discuss the impact of spectrum partition by SH.

5.3.1 Users' Demand Strategies

For the ease of analysis in determine the optimal strategy, we make an approximation:

$$U_{ij}^u \approx \alpha_i w_j^i \ln \frac{g_j^i}{w_j^i} - p_i w_j^i. \tag{5.4}$$

Note that W_i announces the price p_i and it is guaranteed that all the users' demands will be satisfied. Each user optimizes his utility by selecting his optimal demand strategy.

Take the first and second order derivatives, we have:

$$\frac{\partial U_{ij}^u}{\partial w_j^i} = \alpha_i(\ln \frac{g_j^i}{w_j^i} - 1) - p_i. \tag{5.5}$$

and

$$\frac{\partial^2 U_{ij}^u}{\partial {w_j^i}^2} = -\frac{\alpha_i}{w_j^i} < 0. \tag{5.6}$$

So the optimal demand strategy is

$$w_j^{i*} = g_j^i e^{-1-\frac{p_i}{\alpha_i}}. \tag{5.7}$$

As long as $\frac{p_i}{\alpha_i}$ is large enough, $\frac{g_j^i}{w_j^i} \gg 1$ and the error introduced by approximation is very small.

5.3.2 WSPs' Pricing Strategies

In this stage, W_i decides the optimal price p_i based on the auction result and evaluation of users' demands.

Case (1): if $K_i B \leq \sum_{j=1}^{N_i} w_j^{i*}$, the optimal $p_i^* \geq \alpha_i$ because $p_i \sum_{j=1}^{N_i} w_j^{i*} - \sum_{j=1}^{K_i} \tilde{b}_{ij}$ is monotonously decreasing and $p_i K_i B - \sum_{j=1}^{K_i} \tilde{b}_{ij}$ in the region $p_i \in [1, \infty)$. The highest utility is achieved when $\sum_{j=1}^{N_i} w_j^{i*} = K_i B$. We can obtain $p_i^* = \alpha_i(\ln \frac{G_i}{K_i B} - 1)$, where $G_i e^{-1-\frac{p_i}{\alpha_i}} = \sum_{j=1}^{N_i} w_j^{i*}$ is the aggregate demand of W_i's users. In this case, if W_i purchases more bandwidth, it can lower the price and increase the utility.

Case (2): if $K_i B > \sum_{j=1}^{N_i} w_j^{i*}$, Eq. (5.2) can be simplified as

$$p_i \cdot \sum_{j=1}^{N_i} w_j^{i*} - \sum_{j=1}^{K_i} \tilde{b}_{ij}, \tag{5.8}$$

which has the only variable p_i. It can be optimized using numerical methods. To make the analysis easier, we assume $\alpha_i = \alpha_{ij}$ for all j. The unique solution $p_i^* = \alpha_i$ can be obtained. Physically, it means that if W_i has purchased abundant bandwidth, the optimal price that brings highest revenue is $p_i^* = \alpha_i$. Though part of its spectrum is unused, it is nonprofitable to offer a reduced price to stimulate more spectrum demand.

The optimal price is

$$p_i^* = \begin{cases} \alpha_i & \text{if } K_i B > G_i e^{-2} \\ \alpha_i (\ln \dfrac{G_i}{K_i B} - 1) & \text{otherwise} \end{cases} \tag{5.9}$$

5.3.3 WSPs' Bidding Strategies

In this section, we analyze WSPs' bidding strategies in the auction. We take truthfulness and efficiency as the most desirable properties of an auction design.

In our scenario, a WSP indeed sells these spectrums to users, which makes the quantity and evaluation dynamic and flexible. In such case, the challenge is how to determine the best bidding strategies in terms of quantity and evaluation. We propose to decide the best strategies by the pricing scheme and users' demand responses.

Let us start from a simple case. Suppose W_i gets only one channel in the auction. His utility can be calculated by Eq. (5.1) as $U_i(K_i) = U_i(1)$. When W_i considers his bidding strategy, he cannot predict his payment $\{\tilde{b}_{ij}\}$ and K_i.

First, we need to elaborate the true value in this problem. If a buyer is asked to pay the true value to win the object, his utility is zero. In this work, we propose that the true value should be decided according to particular scenarios. We also define the true value as the one that makes a buyer's utility zero given he selects the best strategies in the later stages. So we get W_i's true value for his first channel as:

$$b_{i1}^* = p_i^*(1) \cdot \min\{\sum_{j=1}^{N_i} w_j^{i*}(1), B\}. \tag{5.10}$$

$p_i^*(k)$ and $w_j^{i*}(k)$ stand for the optimal pricing strategy and demand strategy given W_i has k channels. Equation (5.10) is also the marginal benefit that the first channel can bring to W_i. If W_i pays more than this amount to get it, W_i definitely makes a loss.

Similarly, considering the marginal benefit of each additional channel for W_i, the true value for W_i's k-th channel ($2 \le k \le C$) will be

$$\begin{aligned} b_{ik}^* =& p_i^*(k) \cdot \min\{\sum_{j=1}^{N_i} w_j^{i*}(k), kB\} \\ &- p_i^*(k-1) \cdot \min\{\sum_{j=1}^{N_i} w_j^{i*}(k-1), (k-1)B\}. \end{aligned} \tag{5.11}$$

By intuition, the optimal bids' values for any W_i are decreasing as the marginal benefit of the first several channels is higher than the latter ones. We can derive this relationship from Eqs. (5.7) and (5.9)–(5.11) to verify its correctness. We have the theorem:

Theorem 5.1. *Any W_i's bidding structure presents marginal decreasing property. Mathematically, we have,*

$$b_{i1}^* \geq b_{i2}^* \geq \cdots \geq b_{iC}^*. \tag{5.12}$$

Proof. By Eqs. (5.10) and (5.11), we have

$$\sum_{k=1}^{K_i} b_{ik}^* = p_i^*(K_i) \cdot \min\{\sum_{j=1}^{N_i} w_j^{i*}(K_i), K_i B\}. \tag{5.13}$$

Substitute Eq. (5.9) into it, then we get

$$\sum_{k=1}^{K_i} b_{ik}^* = \min\{(\ln(\frac{G_i}{K_i B}) - 1)K_i B, G_i e^{-2}\}. \tag{5.14}$$

Define function

$$f(K_i) = (\ln(\frac{G_i}{K_i B}) - 1)K_i B \tag{5.15}$$

Because

$$\frac{\partial f(K_i)}{\partial K_i} = B(\ln(\frac{G_i}{K_i B}) - 2) \geq 0 \tag{5.16}$$

and

$$\frac{\partial^2 f(K_i)}{\partial K_i^2} = -\frac{B}{K_i} < 0. \tag{5.17}$$

$G_i e^{-2}$ is a constant and the operator min keeps the convexity. So Eq. (5.13) is a convex function.

Note that $f(K_i)$ is positive for all $K_i = 1, 2, \cdots$. By $f(K_i)$'s continuity, we have

$$\begin{aligned} f(0) &= \lim_{K_i \to 0+} f(K_i) \\ &= \lim_{x=\frac{1}{K_i} \to +\infty} \frac{B(\ln(\frac{GX}{B}) - 1)}{x} \\ &= \lim_{x \to +\infty} \frac{B}{x} \\ &= 0. \end{aligned} \tag{5.18}$$

Algorithm 6 Flexible Auction: winner determination

1: Find the C largest bids from $\{b_{ij}\}$, $(i = 1, \cdots, N, j = 1, \cdots, C)$, with a max heap.
2: Let $\{b_i^s\}, i = 1, \cdots, C$ be the sorted array of the C largest bids in descending order.
3: $K_i = 0$, for all $i = 1, \cdots, N$.
4: **for** $k = 1$ to C **do**
5: $K_i = K_i + 1$, where b_k^s is submitted by W_i.
6: **end for**
7: return (W_i wins K_i channels, $i = 1, \cdots, N$).

By Eq. (5.13)'s convexity, we have

$$2\sum_{k=1}^{K_i+1} b_{ik}^* \geq \sum_{k=1}^{K_i} b_{ik}^* + \sum_{k=1}^{K_i+2} b_{ik}^* \tag{5.19}$$

for all $K_i = 0, 1, 2, \cdots$. After simplification, we obtain $b_{i1}^* \geq b_{i2}^* \geq \cdots \geq b_{iC}^*$.

5.3.4 *Auction Design*

There are two goals in the auction design First, the auction mechanism should be truthful. Therefore, W_i bidding with his true values is the dominant strategy. Second, the auction should provide efficiency, which means the channels should be allocated to those bidders who evaluate them most. Therefore, SH's revenue can be increased.

The auction is a sealed-bid auction with one seller (SH) and multiple buyers (WSPs). The auction procedure consists of two parts: the winner determination (channel allocation) and payment mechanism. As the channels are identical, the channel allocation result is presented in the form of the number of winning channels $\{K_i\}$. The payment mechanism can be flexible. In this work, we consider three different payment mechanisms. The first one is the well-known VCG mechanism [2, 5, 10]. The second one is a modified version of the uniform pricing mechanism which preserve truthfulness for multi-unit demands. Besides, we also design a partial uniform pricing mechanism. The advantage of VCG mechanism is that it generates the highest revenue for seller in all truthful auction mechanisms.

5.3.4.1 Winner Determination

The auction is a standard auction such that the C bids with highest values are selected as winning bids: $\sum_{i=1}^{N} K_i = C$. This can be easily achieved by selecting the largest C bids from all NC bids, which is done by building a maximum heap (Algorithm 6).

Algorithm 7 Flexible Auction: payment mechanism (VCG)

```
1: Double the size of {b_i^s} to 2C.
2: Find the 2C largest bids from {b_ij}, (i = 1, ··· , N, j = 1, ··· , C), with a max heap.
3: Store the sorted array of the 2C largest bids in descending order in {b_i^s}, i = 1, ··· , 2C.
4: for i = 1 to N do
5:    if K_i > 0 then
6:       losingBid = 0
7:       j = C + 1
8:       totalPayment(i) = 0
9:       while losingBid < K_i do
10:         if b_j^s is not submitted by W_i then
11:            totalPayment(i) = totalPayment(i) + b_j^s
12:            losingBid = losingBid + 1
13:         end if
14:         j = j + 1
15:      end while
16:   end if
17: end for
18: return (W_i pays totalPayment(i), i = 1, ··· , N).
```

5.3.4.2 Payment Mechanism

The payment mechanism is relatively independent of the previous winner determination part. Here we introduce three possible payment mechanisms, namely: (1) VCG mechanism, (2) modified uniform pricing mechanism, and (3) partial uniform pricing mechanism. The algorithms for payment mechanisms are following the notations in Algorithm 6.

The idea of VCG mechanism is highly abstract and can be applied to universal cases, independent of the form of bidding structure, items to be sold, and so on. With VCG mechanism, W_i's payment is determined by the externality he exerts on other competing WSPs. In this auction, the externality is the sum of other WSPs' K_i highest losing bids. The VCG mechanism is designed as Algorithm 7. Still the same, SH needs to know the highest $2C$ bids only. Lines 1–5 do this job. Because even in the extreme case where one WSP wins all C channels, his externality is the following C largest bids. That means knowledge of largest $2C$ bids is enough.

The general idea of uniform pricing is to charge each channel the same price as the channels are identical. It increases the buyers' acceptance since there is no price discrimination. The uniform price charged is also called the market clearing price. However, in general, uniform pricing is not truthful for multi-unit demand [1]. To guarantee the truthfulness, the clearing price should be selected independent of the winners' bids. Here we introduce a modification of the traditional uniform pricing scheme. Instead of choose the highest losing bids or the lowest winning bids as the clearing price, we use the highest bids from the bidder who loses all his bids. This modified version of uniform pricing only works under the condition that $C < N$. When $C < N$, we can then find a WSP who does not win a channel at all.

Algorithm 8 Flexible Auction: payment mechanism (Uniform pricing)

```
(Given C < N)
maxLoserBid = 0 // Store the highest bid from a bidder who does not win any channel
for i = 1 to N do
    if b_i1 < b^s_C and b_i1 > maxLoserBid then
        maxLosrBid = b_i1
    end if
end for
return (W_i pays totalPayment(i) = maxLosrBid * K_i, i = 1, ··· , N).
```

The modified uniform pricing algorithm is given by Algorithm 8. In the rest of this chapter, we mean "uniform pricing" by the modified version.

Motivated by the previous two mechanisms, we design the partial uniform pricing, which preserves both their advantages. By partial uniform pricing, a winning WSP pays for each of his channels the same amount of money. His unit price is determined by others' highest losing bid. But different WSPs can have different unit prices. There are two advantages of partial uniform pricing compared with uniform pricing. On one hand, this mechanism will generate revenue for SH no less than that of VCG mechanism. An intuitive explanation is given here. By partial uniform pricing mechanism, W_i pays the amount of K_i multiples others' highest losing bid. By VCG mechanism, W_i pays the amount of the sum of others' highest K_i bids. On the other hand, this mechanism does not require the condition $C < N$ since we always have a losing bid as W_i's clearing price. The partial uniform pricing mechanism is shown in Algorithm 9.

We can combine any of the three payment mechanisms with the standard winner determination part to get a complete auction rule. The auction is truthful with any of the three mechanisms. The auction also maximizes social welfare. These two property will be proved later.

5.3.5 Spectrum Partition

From the previous part, we show that the payment mechanism can affect SH's revenue. Not surprisingly, the value of C also does. A significant and interest question is: if SH has a whole spectrum block initially, how many channels should he partition it into in order to maximize his own revenue $U_{SH}(C)$ or social welfare $S(C)$? Note that the size of spectrum block BC is constant.

For the question of maximization of SH's own revenue, there seems no beautiful answer. Intuitively C should not be too small or too large. A small C means that SH sells bigger piece of channels to fewer WSPs. Considering that WSPs' marginal evaluations on spectrum bandwidth are decreasing, it may be good for SH to further divide the big spectrum piece to smaller ones and sell them to more WSPs.

Algorithm 9 Flexible Auction: payment mechanism (Partial uniform pricing)

for $i = 1$ to N **do**
  if $K_j < C$ **then**
    $maxLoserBid(i) = b_{j(K_j+1)}$
  else
    $maxLoserBid(i) = 0$
  end if
end for
Find the largest two elements from the array $maxLoserBid$. Let them be $maxLoserBid1$ and $maxLoserBid2$.
for $i = 1$ to N **do**
  if $maxLoserBid(i) < maxLoserBid1$ **then**
    $totalPayment(i) = maxLoserBid1 \times K_i$
  else
    $totalPayment(i) = maxLoserBid2 \times K_i$
  end if
end for
return (W_i pays $totalPayment(i), i = 1, \cdots, N$).

A large C means there are lots of bids presented, which increases the burden of buyers and incurs high computational overheads.

To maximize the social welfare, we can take any of the three payment mechanisms. Because the affects of payment can be canceled out by the summation of WSPs' and SH's utilities. Maximization of social welfare is equivalent of allocating the channels to those who evaluate them highest, which is exactly what we have done in the winner determination mechanism.

Besides, C also affects the social welfare. If $C \to \infty$, it is a standard water filling problem similar as power control in multi-antenna Gaussian channels (i.e., [7]). There is a well-known algorithm which runs iteratively to obtain the numerical solution. The optimal spectrum partition is a water filling problem with the constraint of granularity of C. So we can conclude that the social welfare increases with C generally. Strictly, we have the property that $S(kC) \geq S(C)$ under the same auction rule, where k is any positive integer. Because after the further partition of current channels, SH can allocate k times of number of channels to current winning WSPs, which leads to the same social welfare. So the further partition of channels does not decrease social welfare. Due to the marginal effect of WSPs' evaluations, the improvement of social welfare by further partition gets smaller while the overhead gets larger. SH can select an appropriate C by considering all factors.

To make a summary, there is no closed-form solution for optimal C to maximize SH's revenue or social welfare though it does exist. When knowing possible service positioning strategies of the WSPs, SH can jointly optimize the decision of C and the auction design to obtain the highest revenue. In practice, SH can select best C according to his purpose, computational overheads and other concerns, such as the technical requirements of wireless communication standards and channel aggregation cost. If circumstances allowed, one simple approach can be that SH iteratively doubles C and stops when $|\frac{S(2C)-S(C)}{S(C)}|$ is less than a predefined threshold. We will leave the detailed discussions as our future work.

5.4 Economic Properties and Time Complexity

In this section, we prove the two properties of Flexauc: truthfulness and efficiency. We also analyze its time complexity. Individual Rationality and Budget Balance is self-evident through the design process and analysis of our algorithms. So we do not discuss these two properties here.

5.4.1 Truthfulness

Theorem 5.2. *Flexauc is truthful with any of the three payment mechanisms such that W_i's best bidding strategy is $\{b^*_{i1}, \cdots, b^*_{ik}, \cdots, b^*_{iC}\}$.*

Proof. To prove the truthfulness, we need to show that for any $1 \le i \le N$ and $1 \le k \le C$, $\{b^*_{i1}, \cdots, b^*_{ik}, \cdots, b^*_{iC}\}$ weakly dominates any other $\{b_{i1}, \cdots, b_{ik}, \cdots, b_{iC}\}$.

First we show that by replacing only b_{ik} with b^*_{ik} (if $b_{i(k-1)} \le b^*_{ik} \le b_{i(k+1)}$), definitely the new strategy $\{b_{i1}, \cdots, b^*_{ik}, \cdots, b_{iC}\}$ weakly dominates the original strategy $\{b_{i1}, \cdots, b_{ik}, \cdots, b_{iC}\}$. By original strategy W_i wins k_i channels and by new one he wins k^*_i channels. We discuss the three possible cases:

1. Case 1: $k_i = k^*_i$. W_i wins the same number of channels by both strategies. By any of the three mechanisms, his payment is determined by other WSPs' bids. So W_i utilities are the same by both strategies.
2. Case 2: $k_i < k^*_i$. It means W_i wins more channel by new strategy. His payment will be no more than the marginal benefit of the additional channel(s). So his utility is improved or keep the same by the new strategy. New strategy dominates original one.
3. Case 3: $k_i > k^*_i$. By new strategies, W_i wins less channels. W_i's bid on k-th channel does matter. b_{ik} is one of the highest C biddings but b^*_{iK} is not. As both b_{ik} and b^*_{ik} are larger than b_{ij} ($j = k+1, k+2, \cdots, C$), so b_{ij} is not one of the highest C biddings. So $k_i = k^*_i + 1$. That means the only difference is that original strategy gets k-th channel and new strategy does not. According to the payment mechanisms, by original strategy, W_i pays more than the marginal benefit b^*_{iK} to get the k-th channel. So New strategy dominates original one.

Then for any strategy $\{b_{i1}, \cdots, b_{iC}\}$, we can adjust it in reverse order repeatedly like bubble sort algorithm: $\{b_{i1}, \cdots, b_{iC}\} \rightarrow \{b_{i1}, \cdots, \min\{b^*_{iC}, b_{i(C-1)}\}\} \rightarrow \{b_{i1}, \cdots, \min\{b^*_{ik}, b_{i(k-1)}\}, \cdots\} \rightarrow \cdots \rightarrow \{b^*_{i1,\ldots}\} \rightarrow \cdots \rightarrow \{b^*_{i1}, b^*_{i2} \cdots\} \rightarrow \cdots$ until it becomes $\{b^*_{i1}, \cdots, b^*_{iC}\}$. It can be achieved by no more than $\frac{C(C+1)}{2}$ adjustments. During the adjustments, the new strategies dominant the old ones. So n$\{b^*_{i1}, \cdots, b^*_{iC}\}$ dominates any $\{b_{i1}, \cdots, b_{iC}\}$.

5.4.2 Efficiency

Theorem 5.3. *Flexauc maximizes social welfare.*

Proof. An auction rule $\mathbf{R}^*$ is efficient if it maximizes social welfare,

$$\mathbf{R}^*(\mathbf{x}) \in arg \max_{\forall \mathbf{R}} \sum E_j x_j, \tag{5.20}$$

where $x_j = 0, 1$. By VCG mechanism, the payment of bidder i is

$$P_i = W(0, \mathbf{x}_{-i}) - W_{-i}(\mathbf{x}). \tag{5.21}$$

Bidder i's utility is

$$\begin{aligned} E_i - P_i &= E_i + W_{-i}(\mathbf{x}) - W(0, \mathbf{x}_{-i}) \\ &= W(\mathbf{x}) - W(0, \mathbf{x}_{-i}). \end{aligned} \tag{5.22}$$

So maximization of his own utility is equivalent to maximization of social welfare.

We know that the uniform pricing auction and partial uniform pricing auction distinguish from VCG auction only in the payment part. The payment effect can be canceled out by the summation of WSPs' and SH's utilities. So auctions with the two payment mechanism also maximize social welfare. It means an auction maximizes social welfare as long as it allocates items to those who evaluate them most. By motivating WSPs to bid truthfully and selecting the highest C bids, Flexauc indeed maximizes social welfare.

5.4.3 Time Complexity

We now analyze the running time of the algorithms.

For Algorithm 6, building heap takes $O(N)$ time. Heap adjustment takes $O(logN)$ time. The complexity of Algorithm 6 is $O(ClogN)$. For Algorithm 7, lines 1–5 take $O(ClogN)$ time. Lines 6–19 take $O(NC)$ time. So the complexity of Algorithm 7 is $O(CN)$. Both Algorithms 8 and 9 take $O(N)$ time.

It means that Flexauc with three payment mechanisms induces the computational overhead $O(NC)$, $O(N + ClogN)$ and $O(N + ClogN)$ respectively. Considering that SH receives in total NC bids from WSPs, the time overhead is definitely acceptable.

5.5 Evaluation

In this part we present simulation results to verify our conclusions, evaluate the performance and compare it with existing mechanisms. The experiment environment is MATLAB. We evaluate WSPs' bidding strategies and show its marginal effect and truthfulness, then WSPs' optimal pricing strategies. The comparison of three payment mechanism and influence of C are also presented. We also compare our mechanism with previous auction mechanism. A frequently used scheme restricts each WSP to submit only one bid and SH to make a 0/1 allocation. We call it OneBid auction. We show that Flexauc always outperforms OneBid. All results have been averaged over abundant of cases with randomly generated parameters.

The default setting of parameters are $C = 5$, $B = 50\,(\text{MHz})$, $N = 10$, N_i is randomly distributed within $[500, 1{,}000]$ for $i = 1, \cdots, N$. $\{\alpha_i\}$ are equally distributed in $[0.2, 0.4]$. The transmission range to the base station is randomly chosen in $[500, 1{,}000]$ (meter) for any user with a uniform distribution. Assume that 75 % of users are indoor and their service requirement is originated from indoor environment, and 25 % from outdoor environment.

We define the attenuation factors as the multiplicative inverses of the path-losses based on the ITU and COST models [9].

1. From outdoor user U_{ij} to base station:

$$H_j = 10^{-4.9}(\frac{r}{1{,}000})^{-4} f^{-3} 10^{-S/10};$$

2. From indoor user U_{ij} to base station:

$$H_j = 10^{-3.7}(\frac{r}{1{,}000})^{-3} 10^{-S/10} 10^{-\frac{18.3n(\frac{n+2}{n+1}-0.46)}{10}};$$

The other default values of parameters are as follows: $P_m = 1$ (watt), $n_0 = -204\,(\text{dB/Hz})$, r (in meters) is the transmission range, $f = 2{,}000\,(\text{MHz})$ is the carrier frequency, $n = 20$ is the number of floors in the path, S is the log-normal shadowing factor with the standard deviation of 8 (dB).

We generate 100 cases with random radio parameters and randomly selected payment mechanism. In each case, we do 100 times of random selection of one WSP (either winner or loser) and make a random adjustment of his true bidding values (while still keeping the marginal effect of bids). Table 5.1 shows the truthfulness

Table 5.1 Truthfulness of WSPs' bids

N,C	$U_i(u) < U_i(t)$	$U_i(u) = U_i(t)$	$U_i(u) > U_i(t)$
10, 5	0.1012	0.8988	0
10, 10	0.1331	0.8669	0
10, 20	0.1721	0.8279	0

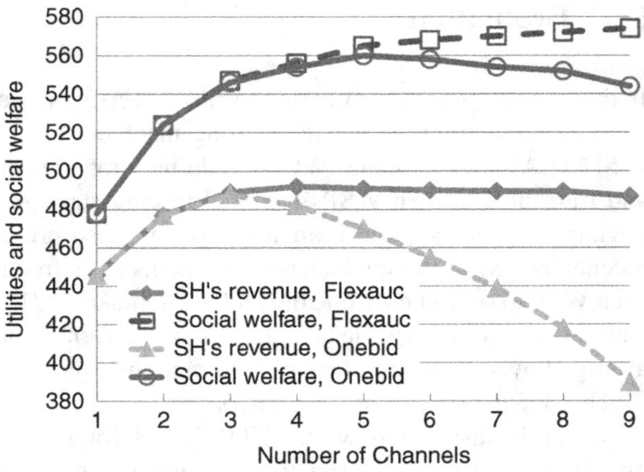

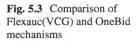

Fig. 5.3 Comparison of Flexauc(VCG) and OneBid mechanisms

of the bids. $U_i(u)$ and $U_i(t)$ are WSP i's utilities when he bids untruthfully and truthfully respectively. The values are the probabilities of the three cases with different N and C values. We see that in any case, $U_i(u) > U_i(t)$ never happens, which supports the truthfulness of our auction. With more WSPs and channels, the probability that $U_i(u) < U_i(t)$ is rising. It means that in such cases, cheating is more likely to reduce the cheater's utility as the competition becomes severe. No matter the WSP is a winner or loser, he has no incentive to bid untruthfully.

Figure 5.3 compares Flexauc with OneBid auction. In OneBid auction, both VCG and partial uniform pricing degrade to uniform pricing. So we do not have other payment mechanism for comparison under OneBid. We compare OneBid with Flexauc(VCG). We find that when C is small enough ($C \leq 3$), the two auction almost perform the same in both SH's revenue and social welfare. When C is larger, Flexauc(VCG) outperforms OneBid and the gaps keep increasing. If $C \geq N$, each WSP will obtain a channel under OneBid. But it must not happen for two reasons. First, there is no clearing price for this case to guarantee the truthfulness. Second, if $C > N$, definitely there are $C - N$ channels wasted. A rational SH will not make $C > N$ under OneBid auction.

Figure 5.4 presents two WSPs' bids structure. The results consist with our theoretical analysis of diminishing marginal value of obtained channels. In this figure, W_2's first bid value is small than W_1's fourth bid. If W_2 wins one channel, then W_1 must win at least four channels.

Figure 5.5 shows the WSPs' utilities under different pricing strategies. We select the three winning WSPs after the auction and calculated their utilities under the prices of $[0.1p_i^*, 0.2p_i^*, \cdots, 2p_i^*]$. Setting higher or lower prices may lead to their overall losses. It verifies the correctness of WSPs' optimal pricing strategies.

Figure 5.6 gives numerical results for optimal C. We see that the larger C is, the higher social welfare it achieves. When C is large enough, the increment of social welfare gets smaller. If SH concerns more about social welfare, he can select, for example, $C = 30$. We also see that when C is too small ($C \leq 2$), both the

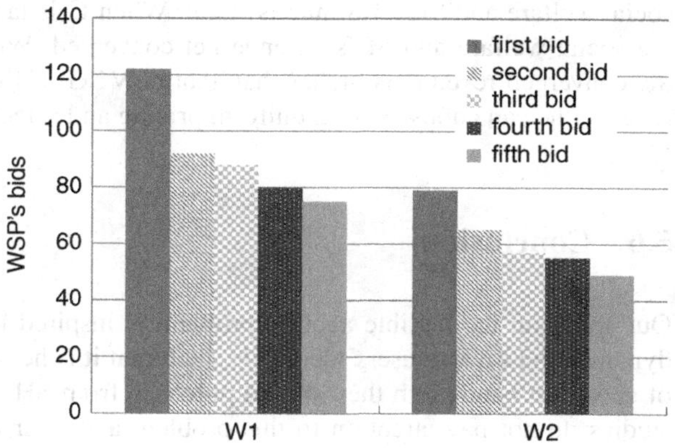

Fig. 5.4 WSPs' bids structure follows the marginal effect

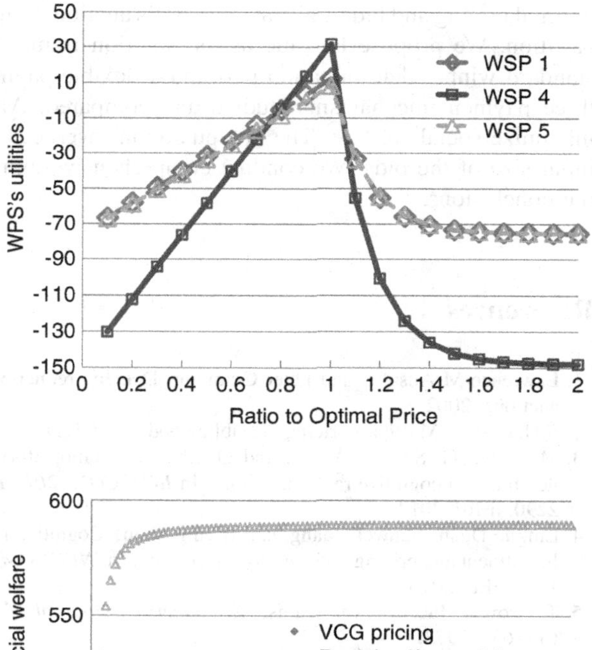

Fig. 5.5 WSPs' optimal pricing strategies

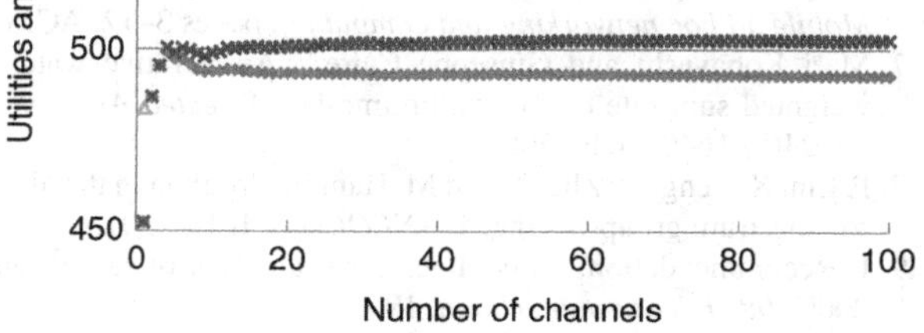

Fig. 5.6 Influences of C on SH's revenue and social welfare

social welfare and SH's revenue is small. When C is large enough ($C \geq 30$), both the social welfare and SH's revenue get converged. With partial uniform pricing, the converged revenue is higher than that of VCG. If SH concerns more about his revenue, he can choose partial uniform pricing and select large enough C.

5.6 Conclusion

Our study on the flexible auction problem is inspired by the fact that WSPs face dynamic and diverse users' demands, such that it is heavily related with how much of spectrum bandwidth they should purchase from SH. Previous spectrum auction studies do not pay attention to this problem and their auction mechanism cannot be directly applied in this scenario or with unsatisfying performance. We make theoretical analysis in the bottom-up manner for users demands response, WSPs' optimal pricing and bidding strategies, SH's auction design and discuss the spectrum partition. We propose Flexauc as the solution framework. Flexauc consists of a standard winner determination part and a flexible payment mechanism. There are three payment mechanisms studied and compared. All of them are truthful and maximize social welfare. The computational overhead of Flexauc is linear to the input size of the bids. We conduct comprehensive numerical simulation to verify our conclusions.

References

1. Lawrence M Ausubel and Peter Cramton. Demand reduction and inefficiency in multi-unit auctions. 2002.
2. E.H. Clarke. Multipart pricing of public goods. *Public choice*, 11(1):17–33, 1971.
3. M. Dong, G. Sun, X. Wang, and Q. Zhang. Combinatorial auction with time-frequency flexibility in cognitive radio networks. In *INFOCOM, 2012 Proceedings IEEE*, pages 2282–2290. IEEE, 2012.
4. Lingjie Duan, Jianwei Huang, and Biying Shou. Cognitive mobile virtual network operator: Investment and pricing with supply uncertainty. In *INFOCOM, 2010 Proceedings IEEE*, pages 1–9. IEEE, 2010.
5. T. Groves. Incentives in teams. *Econometrica: Journal of the Econometric Society*, pages 617–631, 1973.
6. J. Jia, Q. Zhang, Q. Zhang, and M. Liu. Revenue generation for truthful spectrum auction in dynamic spectrum access. In *Proceedings of the tenth ACM international symposium on Mobile ad hoc networking and computing*, pages 3–12. ACM, 2009.
7. Mari Kobayashi and Giuseppe Caire. An iterative water-filling algorithm for maximum weighted sum-rate of gaussian mimo-bc. *Selected Areas in Communications, IEEE Journal on*, 24(8):1640–1646, 2006.
8. P. Lin, X. Feng, Q. Zhang, and M. Hamdi. Groupon in the air: A three-stage auction framework for spectrum group-buying. In *INFOCOM*. IEEE, 2013.
9. I. Recommendation. Guidelines for evaluation of radio transmission technologies for IMT-2000. *Int. Telecommun. Union*, 1997.

10. William Vickrey. Counterspeculation, auctions, and competitive sealed tenders. *The Journal of finance*, 16(1):8–37, 1961.
11. D. Yang, X. Fang, and G. Xue. Truthful auction for cooperative communications. In *Proceedings of the Twelfth ACM International Symposium on Mobile Ad Hoc Networking and Computing*, page 9. ACM, 2011.
12. Youwen Yi, Jin Zhang, Qian Zhang, and Tao Jiang. Spectrum leasing to femto service provider with hybrid access. In *INFOCOM, 2012 Proceedings IEEE*, pages 1215–1223. IEEE, 2012.
13. X. Zhou, S. Gandhi, S. Suri, and H. Zheng. ebay in the sky: strategy-proof wireless spectrum auctions. In *Proceedings of the 14th ACM international conference on Mobile computing and networking*, pages 2–13. ACM, 2008.
14. X. Zhou and H. Zheng. Trust: A general framework for truthful double spectrum auctions. In *INFOCOM 2009, IEEE*, pages 999–1007. IEEE, 2009.

Chapter 6
Conclusions

In this book, we provide an overview of the application of auctions in the wireless communication market. In particular, we introduce detailed designs of three auction schemes in different scenarios from Chaps. 3 to 5.

In the first work, we study the redistribution of heterogeneous channels among multiple sellers and buyers via a double auction mechanism. To overcome the challenges brought by the heterogeneity of the channels in terms of interference range, we design a novel truthful double auction framework called TAHES. TAHES increases spectrum utilization through spectrum reuse. TAHES can not only solve unique challenges caused by spectrum heterogeneity but also preserve nice economic properties: Truthfulness, Budget Balance and Individual Rationality.

In our second work we aim to build a group-buying framework for spectrum trading. In secondary spectrum markets, users in secondary networks may want to bid for spectrum frequencies. But individual users with limited budgets cannot afford the whole spectrum block. Inspired by the emerging group-buying services on the Internet, e.g., Groupon, we propose that users can be voluntarily grouped together to acquire and share the whole spectrum band. We present a three-stage auction-based framework for this problem.

In the third work, we study the spectrum auction in primary market. In wireless markets, major operators buy spectrum through auctions hold by spectrum regulators and serve end users. How much spectrum should an operator buy and how should he set the optimal service tariff to maximize his own benefits are challenging and important research problems. We jointly study the spectrum holder's strategy in the auction and the WSPs' strategies in service provisions. We point out the relationship between their optimal strategies. To meet the WSP's flexible requirements, we design a flexible auction scheme (Flexauc), a novel auction mechanism to enable WSPs to bid for a dynamic number of channels. We prove theoretically that Flexauc not only maximizes the social welfare but also preserves other nice properties: truthfulness and computational tractability.

P. Lin et al., *Auction Design for the Wireless Spectrum Market*, SpringerBriefs in Computer Science, DOI 10.1007/978-3-319-06799-5_6, © The Author(s) 2014

Appendix A
Proof of Theorems in Chapter 5

Proof of Theorem 1

Proof. By Eqs. (5.10) and (5.11), we have

$$\sum_{k=1}^{K_i} b_{ik}^* = p_i^*(K_i) \cdot \min\{\sum_{j=1}^{N_i} w_j^{i*}(K_i), K_i B\}. \tag{A.1}$$

Substitute Eq. (5.9) into it, then we get

$$\sum_{k=1}^{K_i} b_{ik}^* = \min\{(\ln(\frac{G_i}{K_i B}) - 1)K_i B, G_i e^{-2}\}. \tag{A.2}$$

Define function

$$f(K_i) = (\ln(\frac{G_i}{K_i B}) - 1)K_i B \tag{A.3}$$

Because

$$\frac{\partial f(K_i)}{\partial K_i} = B(\ln(\frac{G_i}{K_i B}) - 2) \geq 0 \tag{A.4}$$

and

$$\frac{\partial^2 f(K_i)}{\partial K_i^2} = -\frac{B}{K_i} < 0. \tag{A.5}$$

$G_i e^{-2}$ is a constant and the operator min keeps the convexity. So Eq. (A.1) is a convex function.

P. Lin et al., *Auction Design for the Wireless Spectrum Market*, SpringerBriefs in Computer Science, DOI 10.1007/978-3-319-06799-5, © The Author(s) 2014

Note that $f(K_i)$ is positive for all $K_i = 1, 2, \cdots$. By $f(K_i)$'s continuity, we have

$$\begin{aligned} f(0) &= \lim_{K_i \to 0+} f(K_i) \\ &= \lim_{x=\frac{1}{K_i} \to +\infty} \frac{B(\ln(\frac{GX}{B}) - 1)}{x} \\ &= \lim_{x \to +\infty} \frac{B}{x} \\ &= 0. \end{aligned} \tag{A.6}$$

By Eq. (A.1)'s convexity, we have

$$2 \sum_{k=1}^{K_i+1} b_{ik}^* \geq \sum_{k=1}^{K_i} b_{ik}^* + \sum_{k=1}^{K_i+2} b_{ik}^* \tag{A.7}$$

for all $K_i = 0, 1, 2, \cdots$. After simplification, we obtain $b_{i1}^* \geq b_{i2}^* \geq \cdots \geq b_{iC}^*$.

Proof of Theorem 2

Proof. To prove the truthfulness, we need to show that for any $1 \leq i \leq N$ and $1 \leq k \leq C$, $\{b_{i1}^*, \cdots, b_{ik}^*, \cdots, b_{iC}^*\}$ weakly dominates any other $\{b_{i1}, \cdots, b_{ik}, \cdots, b_{iC}\}$.

First we show that by replacing only b_{ik} with b_{ik}^* (if $b_{i(k-1)} \leq b_{ik}^* \leq b_{i(k+1)}$), definitely the new strategy $\{b_{i1}, \cdots, b_{ik}^*, \cdots, b_{iC}\}$ weakly dominates the original strategy $\{b_{i1}, \cdots, b_{ik}, \cdots, b_{iC}\}$. By original strategy W_i wins k_i channels and by new one he wins k_i^* channels. We discuss the three possible cases:

1. Case 1: $k_i = k_i^*$. W_i wins the same number of channels by both strategies. By any of the three mechanisms, his payment is determined by other WSPs' bids. So W_i utilities are the same by both strategies.
2. Case 2: $k_i < k_i^*$. It means W_i wins more channel by new strategy. His payment will be no more than the marginal benefit of the additional channel(s). So his utility is improved or keep the same by the new strategy. New strategy dominates original one.
3. Case 3: $k_i > k_i^*$. By new strategies, W_i wins less channels. W_i's bid on k-th channel does matter. b_{ik} is one of the highest C biddings but b_{iK}^* is not. As both b_{ik} and b_{ik}^* are larger than b_{ij} ($j = k+1, k+2, \cdots, C$), so b_{ij} is not one of the highest C biddings. So $k_i = k_i^* + 1$. That means the only difference is that original strategy gets k-th channel and new strategy does not. According to the payment mechanisms, by original strategy, W_i pays more than the marginal benefit b_{iK}^* to get the k-th channel. So New strategy dominates original one.

Then for any strategy $\{b_{i1}, \cdots, b_{iC}\}$, we can adjust it in reverse order repeatedly like bubble sort algorithm: $\{b_{i1}, \cdots, b_{iC}\} \rightarrow \{b_{i1}, \cdots, \min\{b^*_{iC}, b_{i(C-1)}\}\} \rightarrow \{b_{i1}, \cdots, \min\{b^*_{ik}, b_{i(k-1)}\}, \cdots\} \rightarrow \cdots \rightarrow \{b^*_{i1,\ldots}\} \rightarrow \cdots \rightarrow \{b^*_{i1}, b^*_{i2} \cdots\} \rightarrow \cdots$ until it becomes $\{b^*_{i1}, \cdots, b^*_{iC}\}$. It can be achieved by no more than $\frac{C(C+1)}{2}$ adjustments. During the adjustments, the new strategies dominant the old ones. So $\{b^*_{i1}, \cdots, b^*_{iC}\}$ dominates any $\{b_{i1}, \cdots, b_{iC}\}$.

Proof of Theorem 3

It is well-known that VCG mechanism maximizes social welfare. We provide a simple proof here.

Proof. An auction rule $\mathbf{R}^*$ is efficient if it maximizes social welfare,

$$\mathbf{R}^*(\mathbf{x}) \in arg \max_{\forall \mathbf{R}} \sum E_j x_j, \tag{A.8}$$

where $x_j = 0, 1$. By VCG mechanism, the payment of bidder i is

$$P_i = W(0, \mathbf{x}_{-i}) - W_{-i}(\mathbf{x}). \tag{A.9}$$

Bidder i's utility is

$$\begin{aligned} E_i - P_i &= E_i + W_{-i}(\mathbf{x}) - W(0, \mathbf{x}_{-i}) \\ &= W(\mathbf{x}) - W(0, \mathbf{x}_{-i}). \end{aligned} \tag{A.10}$$

So maximization of his own utility is equivalent to maximization of social welfare.

We know that the uniform pricing auction and partial uniform pricing auction distinguish from VCG auction only in the payment part. The payment effect can be canceled out by the summation of WSPs' and SH's utilities. So auctions with the two payment mechanism also maximize social welfare. It means an auction maximizes social welfare as long as it allocates items to those who evaluate them most. By motivating WSPs to bid truthfully and selecting the highest C bids, Flexauc indeed maximizes social welfare.